SpringerBriefs in Statistics

More information about this series at http://www.springer.com/series/8921

Paola Lecca

Identifiability and Regression Analysis of Biological Systems Models

Statistical and Mathematical Foundations and R Scripts

 Springer

Paola Lecca
Faculty of Computer Science
Free University of Bozen-Bolzano
Bozen-Bolzano, Italy

ISSN 2191-544X   ISSN 2191-5458   (electronic)
SpringerBriefs in Statistics
ISBN 978-3-030-41254-8   ISBN 978-3-030-41255-5   (eBook)
https://doi.org/10.1007/978-3-030-41255-5

This Springer imprint is published by the registered company Springer Nature Switzerland AG
The registered company address is: Gewerbestrasse 11, 6330 Cham, Switzerland

To my sister, Michela who encouraged me to write this book, and to my parents who have never let me miss their support.

Preface

The book deals with two branches of the inferential statistics: (i) the analysis of identifiability and (ii) the regression analysis. The main applicative domain in which the techniques of model identifiability and the regression analysis are presented is that of the dynamical models in biochemistry and systems biology.

By definition, a model is identifiable if it is theoretically possible to learn the true values of its parameters from an infinite number of observations of it. If a model is identifiable from a given set of experimental data, then there exists a unique set of parameters returning the observed data. Equivalently, if a model is identifiable, different values of its parameters must generate different probability distributions of the observable variables. Regression analysis is a predictive modelling technique, which investigates the causal relationship (expressed as a mathematical model) between a dependent (target) and independent variable(s) (predictor).

Chemical and biological systems of realistic size and complexity often exhibit stiff and non-linear dynamics whose parameter identifiability is not guaranteed and/or for which the most common and most used regression algorithms do not converge. Consequently, biochemical and biological systems are a suitable benchmark for identifiability and regression analysis techniques. A unique solution for the unknown parameters that links any set of inputs to a set of outputs is a critical requirement for any model-based analysis, and, indeed, may become particularly hard for dynamical models of biochemical and, more in general, biological networks. The size of such systems in terms of number of interacting agents, number and type of interactions among them, the stiffness and non-linearity of their dynamics, along with a suboptimal sample size of the experimental observations (due to objective limitations of the experimental investigation on living matter) challenge the identifiability of the putative models. In turn, parameters in the model that are not identifiable pose challenges during the regression analysis, leading to both imprecise parameter estimation and misleading conclusions, and at the end, to the failure of the modelling process.

The book presents the concepts of complexity of a dynamical systems and knowledge inference (Chap. 1); deterministic and stochastic dynamical models, stiff dynamical systems and hybrid stochastic/deterministic simulation algorithms

(Chap. 2); theoretical and algorithmic treatment of observability, identifiability and distinguishability of models of complex systems (Chap. 3); the problem of explicative predictor selection in multi-linear regression, robust regression and non-linear regression (Chap. 4). The book reports also R scripts illustrating the implementation of unsupervised model selection and regression analysis, multi-linear regression, unsupervised model selection and an example of difficult non-linear regression (Chap. 5). At the end of each chapter, some exercises for the self-assessment of the reader are proposed. Although the main application contexts illustrated in the book are the biochemistry and systems biology, the methodologies and the computational techniques described therein are of interest and practical use also in other scientific disciplines.

The book is addressed (i) to university students in the last years of their study courses in scientific disciplines such as chemistry, mathematics and engineering, and physics; (ii) to doctoral students in courses in bioinformatics, bioengineering, systems biology, biophysics, biochemistry, environmental sciences, experimental physics and numerical analysis; and (iii) to researchers, modellers and practitioners in these fields. The prerequisites necessary to understand the contents of the book are the knowledge of the fundamentals of mathematical analysis, probability and statistics that are provided in the first years of university courses in scientific subjects and a basic knowledge of R programming languages.

I am particularly grateful to the colleagues of the Departments of Mathematics of University of Trento for their advices and for their outstanding commitment in didactic and dissemination activities, which will always be a luminous example for me to follow. I am very grateful to the colleagues of the Faculty of Computer Science of the University of Bolzano-Bozen. I thank the colleagues who were able to create a pleasant working environment of cultural exchange and optimal integration of different professional skills, necessary in a hard and delicate process such as writing a book. I thank very much my students of the Universities of Trento and Bolzano-Bozen, as their enthusiasm and their questions have always been for me the motivation and inspiration of my work.

Bozen-Bolzano, Italy   Paola Lecca
November 2019

Contents

Chapter 1
Complex Systems, Data and Inference

Abstract The concepts of complexity and networks are recurrent in modern systems biology. They are intimately linked to the very nature of biological processes governed by mathematically complex laws and orchestrated by thousands of interactions among thousands of molecular components. In this chapter, we explain what it means that a system is complex, what are the mathematical tools and the abstract data structures that we can use to describe a complex system, and finally what challenges the scientific community must face today to deduce a mathematical or computational model from observations experimental.

1.1 The Definition of Complex System: Size, Stiffness and Non-linearity

We usually define a phenomenon as complex when this is an expression of the dynamics of a system made up of many components whose individual behaviour and interactions depend on or are influenced by many factors. The phenomenology of a complex system due to the large number of variables involved and to the relationships that bind these variables is difficult to understand and therefore often scarcely predictable. The experimental investigation of complex systems opens up so as to shed light on a number, even a large number, of variables that describe the components of these systems, and their interactions, but cannot be exhaustive. In particular, for open systems, it is very rare to gain a complete knowledge of the phenomenology and evolution. The mathematical investigation of a complex system from experimental data can, however, in many cases help to identify the possible presence of latent variables or to establish whether the number of variables considered in the experimental investigation is insufficient, but above all can be useful to identify the reasons for the complexity inherent in the interactions between the components of the system in question. A system can be complex because its components are many, because they are highly interacting with each other, because (i) the dynamics of the interactions is not linear, and/or (ii) the system is highly sensitive to initial conditions, and/or (iii) the system is stiff, that is, the values assumed by the variables vary in intervals whose amplitude differs by many orders of magnitude. The mathematics used for the

P. Lecca, *Identifiability and Regression Analysis of Biological Systems Models*,
SpringerBriefs in Statistics, https://doi.org/10.1007/978-3-030-41255-5_1

analysis of complex systems does not necessarily have to be complex mathematics. The most promising way to implement a mathematical treatment that is not too complex and usable by users coming from various disciplines is to organize knowledge on a complex system in a *graph* or in a *hypergraph*. In recent years, biological sciences have made extensive use of graphs and networks to represent complex interacting systems composed of sets of genes, proteins, metabolites and functional chemical compounds of various natures and functions [1–3]. In a graph-like representation, the agents are the vertices and the interactions are indicated by arcs connecting interacting agents. The topology of the graph is usually derived from qualitative and quantitative experimental observations. This type of representation implies a new way of investigating a phenomenology, a way whose perspective is that of the whole system and not just of its individual components. In fact, a system is not just a set of components but is a set of components that interact with each other. The graph that represents it includes both information: those on the components and those on their interactions. The graph also facilitates the construction of a mathematical model of the dynamics of a system since it is a data structure that can be translated into a set of equations or computational procedures. The use of a graph or hypergraph representation not only provides a guide for the construction of the mathematical or computational specification of a model and for the analysis of its properties, but allows to identify the possible controls of its complexity. A sensitivity and robustness analysis of the graph allows to identify the driver nodes of a dynamics, and the cluster of driver nodes of stochastic/deterministic hybrid dynamics due to stiffness.

1.2 Biological Systems as Graphs and Hypergraphs

Frequently, the terms *graph* and *network* are used interchangeably, although their meaning is very different. In this book, we will not give a formal mathematical description of graph and network as in literature there are other good books and articles dealing with it. Here, we would rather highlight, what are the differences between graph and network from the point of view of the processes that we want to represent graphically and mathematically. To mark immediately that graph and net are two different objects, we will use the terms *vertex* and *arc* when we talk about graphs and *node* and *edge* when we talk about the network.

Graphs are combinatorial models representing relationships (arcs) between certain agents (vertices). In biology, the vertices typically describe proteins, metabolites, genes, or other molecular complexes, whereas the arcs represent functional relationships or interactions between the vertices such as "activates" "binds to", "catalyses" or "is converted to" [4]. Furthermore, very often the activation action performed by a vertex is represented as an arc coming out of the vertex and pointing to another arc.

In a graph, every edge connects two nodes, and there are no arcs pointing to other arches. Many biological processes, however, are characterized by more than two participating partners. Klamt et al. [4] bring as an example a metabolic reaction

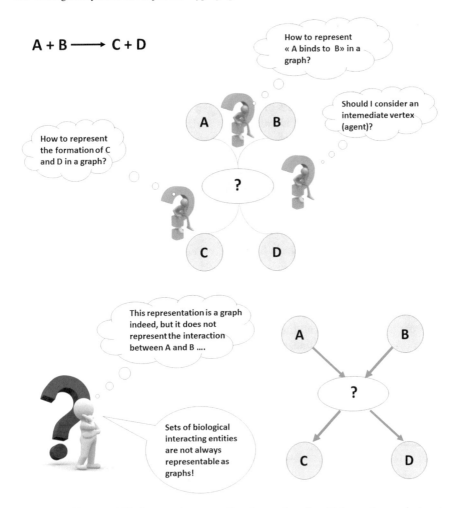

Fig. 1.1 Multireagents' binding reactions, as well as the creation of multiple reaction products are not representable as graphs, where only bilateral relations between nodes are contemplated

involving four species such as $A + B \longrightarrow C + D$, or a protein complex consisting of more than two proteins. Hence, physico-chemical interactions between biological entities are not susceptible of a graph-like representation. As illustrated in Fig. 1.1,[1] an attempt to provide a graph-like representation may cause a loss of information that can lead to wrong interpretations afterwards. A hypergraph is a generalization of a graph that helps to overcome such conceptual limitations [6]. For this reason, many databases and interaction storage formats support hyperedges of different types, either explicitly or implicitly [6–8]. In a hypergraph, an arc can join any number of

---

[1] The clipart objects of "Thinking man" are taken from the free images databases publicly available at free Clipart Library [5].

vertices. What it is commonly called "network" is indeed a hypergraph. Klamt et al. [4] noted that although hypergraphs occur ubiquitously when dealing with cellular networks, their notion is less known than that of graphs. This causes a suboptimal use of the hypergraph expressive potentialities. On the online Encyclopedia of Mathematics [9], we learn that a hypergraf is defined by a set of vertices V and a st od arcs that are defined by subsets of vertices. We learn also that "a hypergraph may be represented in a plane by identifying its nodes with points of the plane and by identifying the edges with connected domains containing the vertices incident with these edges". For example, it is possible to represent a hypergraph H with set of nodes

$$V = \{v_1, v_2, v_3, v_4\}$$

and the family of edges

$$E = \{E_1 = \{v_1\}, E_2 = \{v_2, v_3, v_4\}, E_3 = \{v_2, v_4\}\}.$$

as in Fig. 1.2. A hypergraph H may be also represented by a bipartite graph G as follows: the sets V and E are the partitions of G, and (v_i, E_i) are connected with an edge if and only if vertex v_i is contained in edge E_i in H (Fig. 1.2).

From a set of points and lines on a plane, we can draw a graph with one vertex per point, one vertex per line, and an edge for every incidence between a point and a line. Indeed, a hypergraph is a *incidence structure*. An incidence structure is a triple (P, L, I) where P is a set whose elements are called points, L is a distinct set

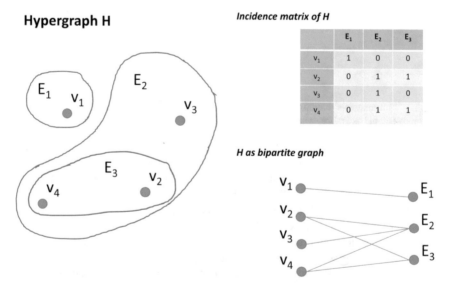

Fig. 1.2 An example of hypergraph. A hypergraph may be represented by a bipartite graph, and conversely

Table 1.1 An example of R script to build and visualize the hypergraph H of Fig. 1.1

```
library(hypergraph)
library(hyperdraw)

nodes <- c("v1", "v2", "v3", "v4", "E1", "E2", "E3")
dh1 <- DirectedHyperedge("v1", "E1")
dh2 <- DirectedHyperedge(c("v2", "v3", "v4"), "E2")
dh3 <- DirectedHyperedge(c("v2", "v4"), "E3")
hg <- Hypergraph(nodes, list(dh1, dh2, dh3))
plot(graphBPH(hg))
```

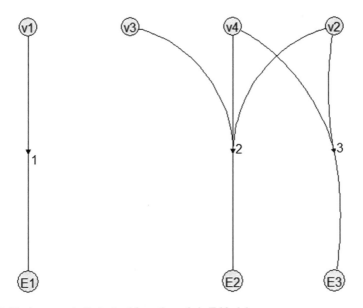

Fig. 1.3 The hypergraph H obtained from the code in Table 1.1

whose elements are called lines and $I \subseteq P \times L$ is the incidence relation. We say that $p \in P$ is in incidence relation with $l \in L$, i.e. $(p, l) \in I$, if p lies on line l. There is a bipartite *incidence graph* or *Levi graph* corresponding to every hypergraph. It is not true, instead that all bipartite graphs can be regarded as incidence graphs of hypergraphs.

In Table 1.1, we show a script in R language [10] for the construction and visualization of the hypergraph H of Fig. 1.1. The result of the code is shown in Fig. 1.3.

In the next section, we will give example of hypergraph models of biochemical and biological interaction networks.

1.2.1 Chemical Reaction and Metabolic Networks

A chemical reaction is a process in which a set of chemical compounds known as *reactants*, $\{R_i\}$, reacts in certain stoichiometric proportions, r_i, to be transformed into a set of other chemical compounds named products, $\{R_i\}$, which are produced in certain stoichiometric quantities p_i:

$$r_1 R_1 + r_2 R_2 + \cdots \longrightarrow p_1 P_1 + p_2 P_2 + \cdots$$

Temkin et al. [11] showed that a chemical reaction can be described as a weighted directed hyperedge in a directed hypergraph where nodes are the chemicals and hyperedges are the reactions. However, Estrada et al. [12] noted that the lack of a mature well-founded theory for the structural analysis of directed hypergraphs caused the co-existence of two alternative commonly used representations of a chemical reaction. In the first representation, a chemical reaction is modelled as a bipartite graph, in which a set of nodes represents the reactants and products and the other set represents the reaction itself. The other representation consists of the *substrate graph*, in which reactants and products are nodes, and two nodes are connected if the corresponding chemical compounds take part in the same reaction. As sets of chemical reactions, metabolic pathways are represented in the form of hypergraphs as well. In order to give an example of metabolic pathway modelled as hypergraph, we consider the amphibolic pathway of the citric acid (Krebs cycle) [13–16], involving the set of reactions reported in Table 1.2, and we present in Table 1.3 the R script to generate a graph-like representation of this pathway, and in Tables 1.4 and 1.5 two R scripts that generate and visualize its hypergraph representation. Although the network of citric acid cycle considered in this example has only 25 reactions and 24 nodes, its graphical representations as hypergraph results to be complex and not immediately understandable, especially compared with the graph representation in Fig. 1.4 (obtained with the R script in Table 1.4). However, the graph is missing important information, for instance, about the citrate formation, that occurs through the reaction

$$\text{acetyl-CoA} + \text{oxaloacetate} + H_2O \longrightarrow \text{citrate} + \text{CoA-SH},$$

which is represented by an edge from acetyl-CoA to oxaloacetate. In many network-like pictures of the citric acid cycle, this reaction is represented as an edge pointing to the edge connecting acetyl-CoA to citrate (Figs. 1.5 and 1.6 and Table 1.6).

1.2.2 Protein Complex Networks

The systematic characterization of multi-protein complexes in the whole proteome of an organism requires the data to be organized in the form of protein membership lists of the protein complexes. The most common forms of this organization are the

Table 1.2 The citric acid cycle, known as *Krebs cycle* is amphibolic. An amphibolic pathway is both anabolic and catabolic in its functions, i.e. it functions in both degradative or catabolic and biosynthetic or anabolic reactions (the Greek prefix "amphi" means "both"). The citric acid cycle is a series of reactions that degrade acetyl Co-enzyme A to yield carbon dioxide, and energy [13–15]

Reaction	Reaction's index
Pyruvate $\longrightarrow$ Acetyl-CoA	R_1
Acetyl-CoA $\longrightarrow$ Oxaloacetate	R_2
Oxaloacetate $\longrightarrow$ Citrate	R_3
Citrate $\longrightarrow$ Cis-aconitate	R_4
Cis-aconitate $\longrightarrow$ Isocitrate	R_5
Isocitrate $\longrightarrow$ Oxalosuccinate	R_6
Oxalosuccinate $\longrightarrow$ alpha-ketoglutarate	R_7
alpha-ketoglutarate $\longrightarrow$ Succinyl-CoA	R_8
Succinyl-CoA $\longrightarrow$ Succinate	R_9
Succinate $\longrightarrow$ Fumarate	R_{10}
Fumarate $\longrightarrow$ Malate	R_{11}
Malate $\longrightarrow$ Oxaloacetate	R_{12}
Malate $\longrightarrow$ Glucose	R_{13}
Citrate $\longrightarrow$ Cholesterol	R_{14}
Citrate $\longrightarrow$ Fatty-Acids	R_{15}
Amino-acids $\longrightarrow$ alpha-ketoglutarate	R_{16}
alpha-ketoglutarate $\longrightarrow$ Amino-acids	R_{17}
Odd_Chains-Fatty-Acids $\longrightarrow$ Succinyl-CoA	R_{18}
Isoleucine $\longrightarrow$ Succinyl-CoA	R_{19}
Methionine $\longrightarrow$ Succinyl-CoA	R_{20}
Valine $\longrightarrow$ Succinyl-CoA	R_{21}
Succinyl-CoA $\longrightarrow$ Porphyrins	R_{22}
Aspartate $\longrightarrow$ Malate	R_{23}
Phenylalanine $\longrightarrow$ Malate	R_{24}
Tyrosine $\longrightarrow$ Malate	R_{25}
Oxalosuccinate $\longrightarrow$ Amino-acids	R_{26}
Amino-acids $\longrightarrow$ Oxalosuccinate	R_{27}

protein–protein interaction networks and the complex intersection graphs. In the first representation, the nodes of the network represent proteins and an edge links two proteins that interact with each other.

Estrada et al. [17] noted that the characterization of multi-protein complexes in the whole proteome of an organism requires that the data are organized in lists of protein membership to protein complexes. This list is usually represented in two ways. The first is the protein–protein interaction network in which the nodes represent proteins and an edge links two proteins that interact with each other. This representation, how-

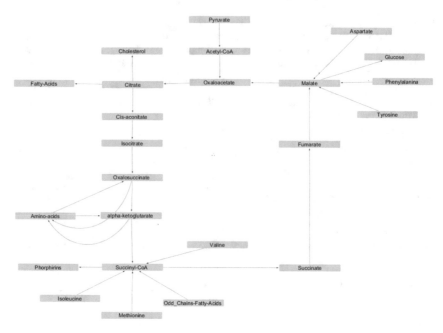

Fig. 1.4 A graph representation of the citric acid cycle (reactions are reported in Table 1.2. The graph has been obtained with the R script in Table 1.2

Table 1.3 A R script to build a graph-like representation of the reactions of citric acid cycle reported in Table 1.2

```
##########################################################
#          The citric acid cycle: the graph             #
##########################################################

require(igraph)

actors <- data.frame(name=unique(c(source, target)))
relations <- data.frame(from=source,
to=target)
g <- graph.data.frame(relations, directed=TRUE, vertices=actors)

# interactive visualization of the graph
tkplot(g, canvas.width = 750, canvas.height = 750, vertex.size=3,
vertex.label.dist=1.5, vertex.color="red", edge.arrow.size=1)

# graph saved in .GRAPHML format (it can be edited with Cytoscape)
write_graph(g, file="Krebs_graph.graphml", format = "graphml")
```

Table 1.4 An example of R script to build and visualize the hypergraph representing the citric acid cycle including the reactions listed in Table 1.2

```
###############################################################
#              The citric acid cycle: the hypergraph          #
#                     - as bipartite graph -                  #
###############################################################

# read the data in table 1.2
net <- read.table("./metabolism.txt", header=T, sep="\t")

source <- as.character(net$Source)
target <- as.character(net$Target)

# build a vector with the names of the reactions (R1, R2, ..., R27)
reaction <- c()
for (i in 1:dim(net)[1])
{
reaction[i] <- paste("R", i, sep="")
}

reagents <-c(unique(c(source, target)), reaction)

dh <- list()
for (i in 1:dim(net)[1])
{
dh[[i]] <- DirectedHyperedge(c(as.character(net[i, 1]),
as.character(net[i, 2])), as.character(net[i, 3]))
}

hg <- Hypergraph(reagents, dh)

plot(graphBPH(hg))
```

ever, does not take into account the multi-protein complexes [17]. The second way is an intersection graph, whose nodes represent complexes, and a link exists between two nodes (complexes) if they have one or more proteins in common. However, the intersection graph does not provide information about proteins. A hypergraph can instead take into account the information about both proteins and common protein membership in the complexes, such as common regulation, localization, turnover or architecture. In the protein complex hypernetworks, each protein is represented by a node and each complex by a hyperedge [17].

Table 1.5 Another example of R script to build and visualize the hypergraph representing the citric acid cycle including the reactions listed in Table 1.2

```
##############################################################
#              The citric acid cycle: the hypergraph           #
##############################################################

krebsgnel <- new("graphNEL",
                nodes=reagents,
                edgeL=list(
                    Pyruvate = list(edges=c("R1")),
                    Acetyl_CoA = list(edges=c("R2")),
                    Oxaloacetate = list(edges=c("R3")),
                    itrate = list(edges=c("R4", "R14", "R15")),
                    s_aconitate = list(edges=c("R5")),
                    Isocitrate = list(edges=c("R6")),
                    Oxalosuccinate = list(edges=c( "R7", "R26")),
                    alpha_ketoglutarate = list(edges=c("R8", "R17")),
                    Succinyl_CoA = list(edges=c("R9", "R22")),
                    Succinate = list(edges=c("R10")),
                    Fumarate = list(edges=c("R11")),
                    Malate = list(edges=c("R12",  "R13")),
                    Amino_acids = list(edges=c( "R16", "R27")),
                    Odd_Chains_Fatty_Acids= list(edges=c("R18")),
                    Isoleucine = list(edges=c("R19")),
                    Methionine = list(edges=c("R20")),
                    Valine = list(edges=c("R21")),
                    Aspartate = list(edges=c("R23")),
                    Phenylalanina = list(edges=c("R24")),
                    Tyrosine = list(edges=c("R25")),
                    Glucose = list(),
                    Cholesterol = list(),
                    Fatty_Acids = list(),
                    Porphirins = list(),
                    #
                    R1 = list(edges= "Acetyl_CoA"),
                    R2 = list(edges= "Oxaloacetate"),
                    R3 = list(edges= "Citrate"),
                    R4 = list(edges= "Cis_aconitate"),
                    R5 = list(edges= "Isocitrate"),
                    R6 = list(edges= "Oxalosuccinate"),
                    R7 = list(edges= "alpha_ketoglutarate"),
                    R8 = list(edges= "Succinyl_CoA"),
                    R9 = list(edges= "Succinate"),
                    R10 = list(edges= "Fumarate"),
                    R11 = list(edges= "Malate"),
                    R12 = list(edges= "Oxaloacetate"),
                    R13 = list(edges= "Glucose") ,
                    R14 = list(edges= "Cholesterol"),
                    R15 = list(edges= "Fatty_Acids"),
                    R16 = list(edges= "alpha_ketoglutarate"),
```

(continued)

Table 1.5 (continued)

```
                    R17 = list(edges= "Amino_acids"),
                    R18 = list(edges= "Succinyl_CoA"),
                    R19 = list(edges= "Succinyl_CoA"),
                    R20 = list(edges= "Succinyl_CoA"),
                    R21 = list(edges= "Succinyl_CoA"),
                    R22 = list(edges= "Porphirins"),
                    R23= list(edges= "Malate"),
                    R24 = list(edges= "Malate"),
                    R25 = list(edges= "Malate"),
                    R26 = list(edges= "Amino_acids"),
                    R27 = list(edges= "Oxalosuccinate")),
                    edgemode = "directed")

krebsbph <- graphBPH(krebsgnel, "^R")
x11(width=600,height=400)
plot(krebsbph)
```

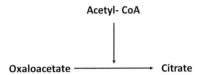

Fig. 1.5 Citrate formation representation often reported in pathways' public databases (e.g. KEGG Pathway database [14], MetaCyc [15])

1.3 Models Construction: Objectives and Challenges

A graph or hypergraph is a specification of the model, which constitutes a bridge between the experimental knowledge of the biological network under examination and the mathematical model described by equations. The graph or hypergraph is constructed from experimental observations and their interpretation. One of the greatest challenges of computational systems biology is the development of mathematical and algorithmic procedures capable of inferring the structure of the graph/hypergraph from experimental data concerning the dynamics and/or equilibrium properties of the system under examination [18, 19]. Even more difficult is the inference of the causal structure of the network that describes the system.

The inference of causal relationships between the hypergraph nodes from experimental data is a non-trivial problem, often referred to as underdetermined one as (i) experimental data does not contain the necessary information to reconstruct the complete network, and/or (ii) the number of variables far exceeds the number of samples. Currently, the problem of underdetermination is becoming of great importance, as

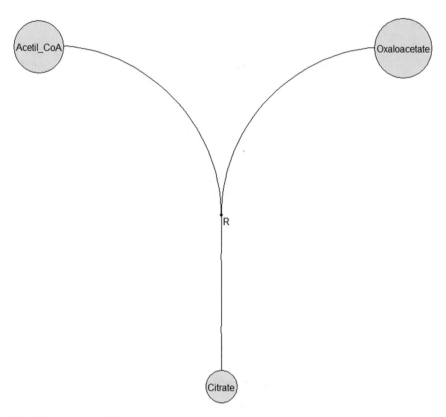

Fig. 1.6 Citrate formation through Acetyl-COA-Oxaloacetate reaction represented as hypergraph. The code in Table 1.5 shows the R script that generates this hypergraph

Table 1.6 A R script to build the hypergraph representing the reactions of citrate formation

```
###############################################################
#    Acetil_CoA + Oxaloacetate --> Citrate: the hypergraph    #
###############################################################

nodes <- c(c("Acetil_CoA", "Oxaloacetate", "Citrate"), "R")
testgnel <- new("graphNEL",
nodes=nodes,
edgeL=list(
Acetil_CoA=list(edges="R"),
Oxaloacetate =list(edges="R"),
Citrate=list(),
R=list(edges="Citrate")),
edgemode="directed")

testbph <- graphBPH(testgnel, "^R")
x11(width=600,height=400)
plot(testbph)
```

omic-size dataset of genes has to be processed to infer new knowledge at the level of system. In this case, underdetermination causes inaccurate results and/or existence of multiple solutions due to the high collinearity of the input data, and, often, also long computational time due to the difficulty of the algorithmic procedures to converge.

Over the last two decades, the identification of gene regulatory networks has been driven by hypothesis-based candidate gene investigation, where a priori knowledge on the topology of the interactions among a subset of genes of interest is considered to facilitate network inference tasks, and keep under control the underdetermination. Most network inference methods developed in the recent past used data generated by low throughput in vivo techniques such as in situ hybridization and immunohistochemistry capable to accurately detect spatio-temporal changes in gene expression in response to targeted gene knock-out, knock-down or over-expression experiments [20]. From the data obtained in many of these experiments, the majority of existent inference methods incrementally identify putative causal gene regulatory networks [21].

An example of challenging inference problem is the gene network inference. It is a still timely topics in computational systems biology, even after a decade of studies in this field. Gene network inference is the process of deduction of gene interactions from experimental data through computational analysis [22]. Since gene networks represent blueprints of cell dynamical processes, reliable inference of large-scale causal gene regulatory interactions can contribute to a better understanding of several aspects of cell physiology, development and pathogenesis. Namely, causal topological structures of the wiring of interactions among genes determine the status (healthy or disrupted) and the collective dynamics of a gene ensemble [23]. In this respect, an efficient and reliable inference procedure can be used to assess the status of the gene network and possibly detect dysregulations in its pathways.

Olsen et al. [24] and Molinelli et al. [25] showed that only in the presence of appropriate sample size and ideal signal-to-noise ratio data methods based on perturbation data it is possible to infer a causal gene regulatory network of n genes in $O(n^2)$ steps. In the last decade, refined methods inspired to perturbation data methods, as nested effect models [26] and methods based on deterministic effects propagation networks [27] attempted to reconstruct causal networks from genome-wide perturbation experiments, even in the presence of unfavourable factors. To what extent computational method can infer a reliable and stable solution from noisy and incomplete non-perturbation data is still an open question. Further, the lack of a proper gold standard makes it difficult to assess gene network method performance. Large-scale community challenges such as the Dialogue for Reverse Engineering Assessments (DREAM) competition [28] have been conducted to assess gene regulatory network inference method either using known regulatory networks in bacteria and yeast or using *in silico* simulated data [29]. While synthetically generated datasets facilitate the performance comparison of a method against a random predictor, the mathematical model used to generate them could potentially favour one or a class of methods over another, thus preventing an objective evaluation of the efficiency of the methods. For instance, while an inference method based on differential equations will likely perform better than a correlation-based approach on synthetic data generated by differential equations, the two approaches might perform equally on real data. Experimental validation would also not fairly assess false positives and false negatives and would be limited to a small fraction of the whole inferred network. Therefore, although a validation on *in silico* benchmark has to be still considered necessary and useful to assess the goodness of a new method, it should be accompanied by a benchmarking against existing approaches, comprehensive tests of output stability, and results should be considered in the light of known relevant data obtained through unrelated sources.

In essence, the inference of networks, especially if gene network, in spite of the availability of many proposed methods, currently suffers from three main limitations: (i) it focuses on the inference of graph (not hypergraph) structures; (ii) it lacks of validation methods that consider not only the topological properties of the graph, but also its response to perturbations and in general the dynamics that the graph wants to represent; and (iii) often, it does not deal with the uncertainty inherent in the reverse engineering process itself. The uncertainty referred to the inference process refers to the fact that the model that describes the data and the interaction processes observed in general is not unique, just as its translation into equations that define its dynamics may not be unique. Interestingly, Rahman et al. [6] proposed using hypergraphs to capture the uncertainty inherent in reverse engineering gene–gene networks. Rahman et al. noted that some subsets of nodes may induce highly varying subgraphs across an ensemble of networks inferred by a reverse engineering algorithm. They claim that a suitable definition of hyperedges can capture this uncertainty in network topology. They implemented an accurate clustering-based approach to identify hyperedges and applied it to a dataset of pathways inferred from genetic interaction data in *S. cerevisiae* related to the unfolded protein response. Unlike the algorithm currently used to compute frequent and dense subgraphs, the Rahman et al. approach discovers several hyperedges that capture the uncertain connectivity of genes in relevant protein complexes. The definition of the problem approached by Rahman et al. is the following:

"Given a set of graphs $\mathscr{G}$, an integer $k > 0$, and parameters $0 < \beta, \sigma \le 1$, enumerate all (β, σ)-hyperedges containing k nodes" [6].

This problem can be formulated as follows. Let (β, σ)-hyperedge S contain k nodes such that $\sigma = 1$, i.e. each of the $2^{\binom{k}{2}}$ possible graphs on S occurs as a subgraph of some graph in $\mathscr{G}$. For such a hyperedge, the largest possible value of β is

$$\beta_{\max} = \frac{1}{2^{\binom{k}{2}}}.$$

As a consequence, since each pair of nodes in S will appear as an edge in precisely half the graphs in $\mathscr{G}$, Rahman et al. concluded that it is possible to compute such a hyperedge by constructing the average of all graphs in $\mathscr{G}$ and searching for cliques in which each edge has weight equal to 0.5.

In conclusion, the inference of a model is a complex process, almost an obstacle course that only an interdisciplinary effort can address. Upstream of each inference procedure is data collection.

Kishan et al. [30] noted that although the topological landscape of gene interaction networks provides a rich source of information for inferring functional patterns of genes or proteins, to aggregate heterogeneous biological information such as gene expression and gene interactions is still an open challenge that is urgent to win to achieve more accurate inference for prediction and discovery of new gene interactions. For this purpose, Kishan et al. proposed a scalable deep learning framework to learn embedded representations to unify known gene interactions and gene expression for gene interaction predictions. The way is thus paved.

The topological landscape of gene interaction networks provides a rich source of information for inferring functional patterns of genes or proteins. However, it is still a challenging task to aggregate heterogeneous biological information such as gene expression and gene interactions to achieve more accurate inference for prediction and discovery of new gene interactions. In particular, how to generate a unified vector representation to integrate diverse input data becomes urgent.

Alongside all the efforts that must be made upstream of the inference, there must be an adequate effort on the part of mathematicians, computer scientists and computational biologists to create data structures and computational models suitable for managing the complexity of the problem. In this chapter, we have talked about *complexity* and we have introduced the concept of hypergraph, a tool that promises to be able to describe this complexity and that still few studies have been carried out, especially in order to analyse its topological properties, which are widely studied for graphs.

1.4 Exercises

Exercise 1.1 Give a hypergraph representation of the Michaelis–Menten reactions system:

$$E + S \rightleftharpoons ES \longrightarrow P,$$

where E, S, ES and P denote, respectively, the enzyme, its substrata, the complex enzyme–substrate and the product.

Exercise 1.2 Give a definition of the concept of a *path* in a directed hypergraph.

Hints: An undirected hypergraph is an extension of the concept of a graph in the sense that edges are allowed to connect an arbitrary number of vertices (instead of exactly two, as in graphs). Hence, a hypergraph is defined as a tuple (V, H), where V is a set of vertices and H is a set of hyperedges, where $H \subset P(H) \, \emptyset$, with $P(H)$ the *power set* (i.e. the set of all possible subsets) of H. A directed hypergraph is an orientation of a hypergraph $H = (V, H)$, i.e. a head node is assigned to every hyperedge in H [31]. Use this definition to give the definition of paths of hyperedges in a directed hypergraph.

Exercise 1.3 Prove that undirected graphs are special cases of hypergraphs in which every hyperedge contains two nodes (i.e. has a cardinality of two).

Exercise 1.4 Write an algorithm to project hypergraph onto graphs. Apply this algorithm to project onto a graph the hypergraph representing the Michaelis–Menten reaction system given in Exercise 1.1. Which information will be lost in this projection?

Exercise 1.5 Give bipartite graph and a hypergraph representation of the following reaction system:

$$R_1 : A + B \longrightarrow C + D$$
$$R_2 : D + E \longrightarrow F + G$$
$$R_3 : F + G \longrightarrow H + I$$
$$R_4 : I \longrightarrow J + K$$
$$R_5 : A + L \longrightarrow C.$$

Following the examples provided in Tables 1.1, 1.4 and 1.5, write a R script to build and visualize the bipartite graph and the hypergraph.

References

1. Ma'ayan A. Complex systems biology. J R Soc Interface. 2017;14(134):20170391.
2. Galas DJ, Sakhanenko NA, Skupin A, Ignac T. Describing the complexity of systems: multivariable "set complexity" and the information basis of systems biology. J Comput Biol. 2014;21(2):118–40.
3. Lesne A. Complex networks: from graph theory to biology. Lett Math Phys. 2006;78(3):235–62.
4. Klamt S, Haus U-U, Theis F. Hypergraphs and cellular networks. PLOS Comput Biol. 2009;5(5):1–6, 05.
5. Clipart Library. http://clipart-library.com/picture-of-thinking-man.html. Accessed 3 Sept 2018.
6. Rahman A, Poirel CL, Badger DJ, Estep C, Murali TM. Reverse engineering molecular hypergraphs. IEEE/ACM Trans Comput Biol Bioinf. 2013;10(5):1113–24.
7. Demir E, Cary MP, Paley S, Fukuda K, Lemer C, Vastrik I, Wu G, D'eustachio P, Schaefer C, Luciano J, Schacherer F, Martinez-Flores I, Hu Z, Jimenez-Jacinto V, Joshi-Tope G, Kandasamy K, Lopez-Fuentes AC, Mi H, Pichler E, Rodchenkov I, Splendiani A, Tkachev S, Zucker J, Gopinath G, Rajasimha H, Ramakrishnan R, Shah I, Syed M, Anwar N, Babur Ö, Blinov M, Brauner E, Corwin D, Donaldson S, Gibbons F, Goldberg R, Hornbeck P, Luna A, Murray-Rust P, Neumann E, Ruebenacker O, Samwald M, van Iersel M, Wimalaratne S, Allen K, Braun B, Whirl-Carrillo M, Cheung K-H, Dahlquist K, Finney A, Gillespie M, Glass E, Gong L, Haw R,

Honig M, Hubaut O, Kane D, Krupa S, Kutmon M, Leonard J, Marks D, Merberg D, Petri V, Pico A, Ravenscroft D, Ren L, Shah N, Sunshine M, Tang R, Whaley R, Letovksy S, Buetow KH, Rzhetsky A, Schachter V, Sobral BS, Dogrusoz U, McWeeney S, Aladjem M, Birney E, Collado-Vides J, Goto S, Hucka M, Le Novère N, Maltsev N, Pandey A, Thomas P, Wingender E, Karp PD, Sander C, Bader GD. The BioPAX community standard for pathway data sharing. Nat Biotechnol. 2010;28(9):935–42.

8. Schaefer CF, Anthony K, Krupa S, Buchoff J, Day M, Hannay T, Buetow KH. PID: the pathway interaction database. Nucleic Acids Res. 2008;37(suppl_1):D674–9.

9. Encyclopedia of Mathematics:. www.encyclopediaofmath.org/index.php/Hypergraph. Accessed 3 Sept 2018.

10. The R Project for Statistical Computing. https://www.r-project.org/. Accessed 10 Jan 2019.

11. Temkin ON, Zeigarnik AV, Bonchev D. Chemical reaction networks: a graph-theoretical approach; 1996.

12. Estrada E, Rodríguez-Velázquez2 JA. Complex networks as hypergraphs; 2005. http://cds.cern.ch/record/836579/files/?ln=it.

13. Stenesh J. The citric acid cycle. In: Biochemistry. Springer US; 1998. p. 273–91.

14. KEGG Pathway Database. https://www.genome.jp/kegg/pathway.html. Accessed 03 Feb 2019.

15. MetaCyc. https://metacyc.org/. Accessed 03 Feb 2019.

16. Wikipdia: metabolic pathways. https://en.wikipedia.org/wiki/Metabolic_pathway. Accessed 03 March 2019.

17. Estrada E, Rodríguez-Velázquez JA. Subgraph centrality and clustering in complex hypernetworks. Phys A: Stat Mech Appl. 2006;364:581–94.

18. Lecca P, Re A, Ihekwaba AE, Mura I, Nguyen T-P. Computational systems biology: inference and modelling. Sawston: Woodhead Publishing; 2016.

19. Oates CJ, Mukherjee S. Network inference and biological dynamics. Ann Appl Stat. 2012;6(3):1209–35.

20. Djordjevic D, Yang A, Zadoorian A, Rungrugeecharoen K, Ho JW. How difficult is inference of mammalian causal gene regulatory networks? PLoS ONE. 2014;9(11):e111661.

21. Davidson EH. Emerging properties of animal gene regulatory networks. Nat. 2010;468:911–920.

22. Äijö T, Bonneau R. Biophysically motivated regulatory network inference: progress and prospects. Human Heredity. 2016;81(2):62–77.

23. Ghersi D, Singh M. Disentangling function from topology to infer the network properties of disease genes. BMC Syst Biol. 2013;7(1):5.

24. Olsen C, Fleming K, Prendergast N, Rubio R, Emmert-Streib F, Bontempi G, Haibe-Kains B, Quackenbush J. Inference and validation of predictive gene networks from biomedical literature and gene expression data. Genomics. 2014;103(5–6):329–36.

25. Molinelli EJ, Korkut A, Wang W, Miller ML, Gauthier NP, Jing X, Kaushik P, He Q, Mills G, Solit DB, Pratilas CA, Weigt M, Braunstein A, Pagnani A, Zecchina R, Sander C. Perturbation biology: inferring signaling networks in cellular systems. PLoS Comput Biol. 2013;9(12):e1003290.

26. Vaske CJ, Benz SC, Sanborn JZ, Earl D, Szeto C, Zhu J, Haussler D, Stuart JM. Inference of patient-specific pathway activities from multi-dimensional cancer genomics data using PARADIGM. Bioinformatics. 2010;26(12):i237–45.

27. Holger F, Özgür S, Dorit A, Christian B, Tim B. Deterministic effects propagation networks for reconstructing protein signaling networks from multiple interventions. BMC Bioinf. 2009;10(1).

28. Hill SM, Heiser LM, Cokelaer T, Unger M, Nesser NK, Carlin DE, Zhang Y, Sokolov A, Paull EO, Wong CK, Graim K, Bivol A, Wang H, Zhu F, Afsari B, Danilova LV, Favorov AV, Lee WS, Taylor D, Hu CW, Long BL, Noren DP, Bisberg AJ, Afsari B, Al-Ouran R, Anton B, Arodz T, Sichani OA, Bagheri N, Berlow N, Bisberg AJ, Bivol A, Bohler A, Bonet J, Bonneau R, Budak G, Bunescu R, Caglar M, Cai B, Cai C, Carlin DE, Carlon A, Chen L, Ciaccio MF, Cokelaer T, Cooper G, Creighton CJ, Daneshmand S-M-H, de la Fuente A, Di Camillo B, Danilova LV, Dutta-Moscato J, Emmett K, Evelo C, Fassia M-KH, Favorov AV, Fertig EJ,

Finkle JD, Finotello F, Friend S, Gao X, Gao J, Garcia-Garcia J, Ghosh S, Giaretta A, Graim K, Gray JW, Großeholz R, Guan Y, Guinney J, Hafemeister C, Hahn O, Haider S, Hase T, Heiser LM, Hill SM, Hodgson J, Hoff B, Hsu CH, Hu CW, Hu Y, Huang X, Jalili M, Jiang X, Kacprowski T, Kaderali L, Kang M, Kannan V, Kellen M, Kikuchi K, Kim D-C, Kitano H, Knapp B, Komatsoulis G, Koeppl H, Krämer A, Kursa MB, Kutmon M, Lee WS, Li Y, Liang X, Liu Z, Liu Y, Long BL, Lu S, Lu X, Manfrini M, Matos MRA, Meerzaman D, Mills GB, Min W, Mukherjee S, Müller CL, Neapolitan RE, Nesser NK, Noren DP, Norman T, Oliva B, Opiyo SO, Pal R, Palinkas A, Paull EO, Planas-Iglesias J, Poglayen D, Qutub AA, Saez-Rodriguez J, Sambo F, Sanavia T, Sharifi-Zarchi A, Slawek J, Sokolov A, Song M, Spellman PT, Streck A, Stolovitzky G, Strunz S, Stuart JM, Taylor D, Tegnér J, Thobe K, Toffolo GM, Trifoglio E, Unger M, Wan Q, Wang H, Welch L, Wong CK, Wu JJ, Xue AY, Yamanaka R, Yan C, Zairis S, Zengerling M, Zenil H, Zhang S, Zhang Y, Zhu F, Zi Z, Mills GB, Gray JW, Kellen M, Norman T, Friend S, Qutub AA, Fertig EJ, Guan Y, Song M, Stuart JM, Spellman PT, Koeppl H, Stolovitzky G, Saez-Rodriguez J, Mukherjee S. Inferring causal molecular networks: empirical assessment through a community-based effort. Nat Methods. 2016;13(4):310–8.
29. Schaffter T, Marbach D, Floreano D. GeneNetWeaver: in silico benchmark generation and performance profiling of network inference methods. Bioinformatics. 2011;27(16):2263–70.
30. Kishan KC, Li R, Cui F, Yu Q, Haake AR. GNE: a deep learning framework for gene network inference by aggregating biological information. BMC Syst Biol. 2019;13(S2).
31. Frank A, Kirly T, Király Z. On the orientation of graphs and hypergraphs. Discret Appl Math. 2003;131(2):385–400. Submodularity.

Chapter 2
Dynamic Models

Abstract A graph or hypergraph is a static representation of all possible interactions between nodes. However, due to these same interactions, the network topology evolves over time. The abundances of the chemical and/or molecular species they represent in nodes change over time and if they fall below a critical threshold, they cause the disappearance of any connected arches and then determine their reappearance if they return to exceed this critical threshold. The description of the dynamics of a network consists of a mathematical model often constituted by differential equations that express the speed of variation of the abundances of the biological entities represented by the nodes. The dynamics of a network can be deterministic, or stochastic or stochastic/deterministic hybrid. Depending on the nature of its determination, the dynamics is modelled by deterministic differential equations, stochastic differential equations, master equations, and in cases where the numerical solution of the latter is difficult to calculate, from stochastic simulation algorithms. In this chapter, we give an overview of the most used dynamic models for simulating the temporal evolution of a network.

2.1 Chemical Networks as Dynamical Systems

From the formal point of view, chemical reaction is handled as a temporal process and a network of chemical interactions is considered a dynamical system [1].

A dynamical system is an ordered pair: (A, ϕ), where A is the state space, and $\phi : T \times A \rightarrow A$ is a function which assigns to the couple (t, x_0), where $x_0 \in A$ is an arbitrary point, the point $x \in A$, that characterizes the state at the time t, assuming that the system was in x_0 at $t = 0$. A fundamental property of ϕ is the validity of the identity

$$\phi((t + s), x_0) = \phi(s, \phi(t, x_0)). \tag{2.1}$$

The motion of a dynamical system is the one variable function

$$\phi_{x_0} : T \rightarrow A \tag{2.2}$$

$$\phi_{x_0} \equiv \phi(\cdot, x_0) \tag{2.3}$$

© The Author(s), under exclusive license to Springer Nature Switzerland AG 2020
P. Lecca, *Identifiability and Regression Analysis of Biological Systems Models*,
SpringerBriefs in Statistics, https://doi.org/10.1007/978-3-030-41255-5_2

where $T \subset \mathbb{R}$ and $A \subset \mathbb{R}^J$, or A consists of random variables taking their values from $\mathbb{R}^M$. For every $t \in T$, $\phi(t, \cdot) : A \to A$ is an automorphism.

The chemical reaction can be classified either by the properties of the *process time*, or by the structure of the *state space*, or by the nature of *determination* [1].

2.1.1   Properties of the Process Time

We report here some considerations about the mathematical model for the time variables from the work of Lecca et al. [1]. The time can be chosen as continuous ($T \subset \mathbb{R}$) or a discrete ($T \subset \mathbb{Z}$) variable. Both the continuous- and the discrete-time models present advantages, disadvantages, arguments in favour and arguments in disfavour. The arguments generally adopted for choosing a continuous-time variable are as follows:

1. Calculation with continuous-time models has greater tradition. Continuous models have the advantage over discrete-time models in that they are more amenable to algebraic manipulation, although they are slightly harder to implement on a computer.
2. Most physical processes are inherently continuous in time. In particular, some physico-chemical quantities can be transduced continuously. Thus, the parameters in the models are strongly correlated with the physical properties of the systems, something that is very appealing to an engineer. Moreover, as cost of computation becomes cheaper, today's data acquisition equipment can provide nearly continuous-time measurements. Fast sampled data can be more naturally dealt with using continuous-time models than discrete-time models.

Arguments for selecting a discrete-time variable are the following.

1. The idea that time has no objective existence but depends on events led some scientists to abandon the assumption that it is a continuous variable. Moreover, we perceive temporal intervals of finite duration rather than durationless instants; and the researcher prefer to assume that the nature has properties that can be verified.
2. The notion of "immediate next time" can be easily interpreted, and this is not so easy in the case of continuous time.
3. The experimentalists measure at discrete points only.

2.1.2   Properties of State Space

The state space can be chosen either continuous or discrete. To emphasize the existence of elementary particles of a population as in reaction kinetics, a discrete state space formalism is preferred.

The notion of *state* was derived from mechanics and thermodynamics and generalized by mathematical system theory. The quantities of a model can be classified into two categories: *state variables* and *constitutive quantities*.

State variables are functions whose values specify the system state, whereas constitutive variables are functions of the state since their values are univocally determined once the state of the system has been assigned. Thus, a constitutive quantity Ω can be expressed as follows:

$$\Omega(t) = \omega(g(t), t), \tag{2.4}$$

where g denotes the state of the system and $\omega : A \times T \to \mathbb{R}'$ is the constitutive functional[1] mapping the state into a constitutive quantity $r \in \mathbb{N}$. The case $r = 1$ means that the value of the constitutive quantity is a scalar. The state of an M-component chemical system is described by a vector [2]:

$$\mathbf{x} : T \to \mathbb{R}^M, \ t \mapsto \mathbf{x}(t) \in \mathbb{R}^M. \tag{2.5}$$

According to the conventional treatment of pure homogeneous reaction Kinetics, the state is a finite-dimensional vector and the only constitutive quantities are the reaction rates.

The theory of thermodynamics adopts the concept of "particles with memory". According to this concept, the constitutive quantities depend on the history of the independent variables and not on their present value. This means that it is not definite that the instantaneous value of state variables (i.e. state) completely determines the state.

Let us introduce the *site* function $h : T \to \mathbb{R}^M$. Since the state is determined by earlier values of the site function, therefore the state g is interpreted as

$$g : T \to G, \ t \mapsto h',$$

where h' is known as *history* function defined as

$$h'(s) = h(t - s), \quad s > 0.$$

Knowing the history, the state can be set up

$$\mathcal{H}(h, \cdot) = g,$$

i.e. $\mathcal{H}$ is a mapping such that

---

[1] The functional assigns a number to a function. Here, the term refers to every mapping having the function as argument.

[2] The statement that a state is described by function is not in contradiction with the statement that a state is a vector. Namely, a finite-dimensional vector can also be interpreted as a function.

$$\mathscr{H}(h, t) = g(t) = \mathscr{H}(h') = h'.$$

If we assume that the history of the site does not influence the state, then the constitutive functional reduces to a function. Furthermore, if we also assume the invertibility of this function, then the differences between the state variables and constitutive quantities are not significant. These two assumptions are tacitly adopted in the classical theories of the thermodynamics.

2.2 Nature of Determination of a Dynamical System

2.2.1 Deterministic Systems

An (A, ϕ) dynamic system is deterministic if knowing the state of the system at one time means that the system is uniquely specified for all $t \in T$. The simplest deterministic model for chemical kinetics is the ordinary differential equation-based model of the law of mass action. This law states that the rate of a chemical reaction, i.e. the measure of how the concentrations of the involved substances changes with time, is defined the Van't Hoff expression

$$k \cdot \Pi_{j=1}^{J} c_j^{s_j}. \tag{2.6}$$

This means that the rate of a reaction is directly proportional to the product of the concentrations of the reactants. k is the specific reaction rate depending on the cross section of collision between the molecules of the reagents and the probability for the collision to result in a reaction. This probability is calculated as the product of the reactants concentrations, $c_j = x_j/V$, raised to the power of their stoichiometry. The Van't Hoff expression in (2.6) gives the number of collision per unit time per unit volume in which $\{x_j\} \longrightarrow \{x_j + s_j - r_j\}$. The rate equations are therefore

$$\frac{dx_i}{dt} = Vk(r_i - s_i)\Pi_{j=1}^{J}\left(\frac{x_j}{V}\right)^{s_j}. \tag{2.7}$$

This equation holds when the following physical requirements are satisfied.

1. The mixture is homogeneous, so that its density equals x_j/V at each point in V.
2. The elastic, non-reactive collisions are sufficiently frequent to ensure that the Maxwell velocity distribution is maintained.
3. The internal degrees of freedom of the molecules are supposed to be in thermal equilibrium, with the same temperature T as the velocities. If not, the fraction of collisions that result in a reaction would depend on the details of the distribution over internal states, and not just on the concentrations.
4. The temperature is uniform in space and constant in time so that the reaction rate coefficients are constant in space and time.

Although these assumptions are only approximations of real scenarios, they do not violate any physical law and their validity can therefore be approximated to any desired accuracy in suitable experiments. They assure that the state of the mixture is fully described by the state vector $\mathbf{x}$. Moreover, these same assumptions are also made in the formulation of stochastic models of chemical kinetics [2].

2.2.2 Stochastic Systems

When the state of the system can be assigned to a set of values with a certain probability distribution, the future behaviour of the system can be determined stochastically. Discrete-time, discrete state space (first order) Markov processes (i.e. Markov *chain*) are defined by the formula

$$\mathscr{P}(\xi_{t+1} = a | \xi_0 = a, \xi_1 = a_1, \ldots, \xi_t = a_t) = \mathscr{P}(\xi_{t+1} = a | \xi_t = a_t), \qquad (2.8)$$

where the set $\{\xi_t | t = 0, 1, 2, \ldots\}$ is a discrete-time stochastic process.

Knowing the total history of the process, we can extrapolate its future behaviour with the same probability as if we knew only the actual current state. The behaviour of the chain is therefore determined by $\mathscr{P}(\xi_{t+1} = a | \xi_t = a_t)$, and thus it depends on a and t. However, if there is no t dependence, so that

$$\mathscr{P}(\xi_s = x | \xi_t = y) = \mathscr{P}_{xy}(s - t),$$

i.e. the transition probabilities are stationary, the Markov chain is said to be *time homogeneous*. In this case, the law of evolution of the system does not depend explicitly on time and consequently, the time origin can be defined arbitrarily. Deterministic dynamic systems generated by ordinary differential equations

$$\frac{dx(t)}{dt} = f(x(t))$$

can be associated with the time homogeneous Markov processes.

A Markov process is a special case of a stochastic process. A stochastic process is a random function $f(X; t)$, where X is a stochastic variable and t is the time. The definition of a stochastic variable consists in specifying

- a set of possible values (called "set of states" or "sample space") and
- a probability distribution over this set.

The set of states may be *discrete*, e.g. the number of molecules of a certain component in a reacting mixture. Alternatively, the set may be *continuous* in a given interval, e.g. one velocity component of a Brownian particle and the kinetic energy of that particle. Finally, the set may be partly discrete and partly continuous, e.g. the energy of an electron in the presence of binding centres. Moreover, the set of states may be *multidimensional*: in this case, X is written as a vector $\mathbf{X}$. Examples: $\mathbf{X}$ may stand for the three velocity components of a Brownian particle or for the collection of all numbers of molecules of the various components in a reacting mixture.

The probability distribution, in the case of a continuous one-dimensional range, is given by a function $P(x)$ that is non-negative

$$P(x) \geq 0$$

and normalized in the sense

$$\int P(x)dx = 1,$$

where the integral extends over the whole range. The probability that X has a value between x and $x + dx$ is

$$P(x)dx.$$

Often in physical and biological sciences, a probability distribution is visualized by an "ensemble". From this point of view, a fictitious set of an arbitrary large number $\mathcal{N}$ of quantities, all having different values in the given range, is introduced. In such a way, the number of these quantities having a value between x and $x + dx$ is $\mathcal{N}P(x)dx$. Thus, the probability distribution is replaced with a density distribution of a large number of "samples". This does not affect any simulation result, since it is merely a convenience in talking about probabilities, and in this work we will use this language.

Finally, we remark that in a continuous range it is possible for $P(x)$ to involve delta functions,

$$P(x) = \sum_n p_n \delta(x - x_n) + \tilde{P}(x),$$

where $\tilde{P}$ is finite or at least integrable and non-negative, $p_n > 0$, and

$$\sum_n p_n + \int \tilde{P}(x)dx = 1.$$

Physically, this may be visualized as a set of discrete states x_n with probability p_n embedded in a continuous range. If $P(x)$ consists of δ functions alone (i.e. $\tilde{P}(x) = 0$), then it can also be considered as a probability distribution p_n on the discrete set of states x_n.

A general way to specify a stochastic process is to define the joint probability densities for values $x_1, x_2, x_3, \ldots$ at times $t_1, t_2, t_3, \ldots$, respectively,

$$p(x_1, t_1; x_2, t_2; x_3, t_3; \ldots). \tag{2.9}$$

If all such probabilities are known, the stochastic process is fully specified (but, in general, it is not an easy task to find all such distributions). Using (2.9), the conditional probabilities can be defined as usual

$$p(x_1, t_1; x_2, t_2; \ldots | y_1, \tau_1; y_2, \tau_2; \ldots) = \frac{p(x_1, t_1; x_2, t_2; \ldots | y_1, \tau_1; y_2, \tau_2; \ldots)}{p(y_1, \tau_1; y_2, \tau_2; \ldots)},$$

where $x_1, x_2, \ldots$ and $y_1, y_2, \ldots$ are values at times $t_1 \geq t_2 \geq \cdots \geq \tau_1 \geq \tau_2 \geq \ldots$. This is where a Markov process has a very attractive property. It has no memory. For a Markov process

$$p(x_1, t_1; x_2, t_2; \ldots | y_1, \tau_1; y_2, \tau_2; \ldots) = p(x_1, t_1; x_2, t_2; \ldots | y_1, \tau_1)$$

the probability to reach a state x_1 at time t_1 and state x_2 at time t_2, if the state is y_1 at time τ_1, is independent of any previous state, with times ordered as before. This property makes it possible to construct any of the probabilities (2.9) by a *transition probability* $p_\rightarrow(x, t | y, \tau)$, $(t \geq \tau)$, and an initial probability distribution $p(x_n, t_n)$:

$$p(x_1, t_1; x_2, t_2; \ldots x_n, t_n) =$$
$$p_\rightarrow(x_1, t_1 | x_2, t_2) p_\rightarrow(x_2, t_2 | x_3, t_3) \ldots$$
$$\ldots p_\rightarrow(x_{n-1} t_{n-1} | x_n, t_n) p(x_n, t_n).$$

A consequence of the Markov property is the Chapman–Kolmogorov equation

$$p_\rightarrow(x_1, t_1 | x_3, t_3) = \int p_\rightarrow(x_1, t_1 | x_2, t_2) p_\rightarrow(x_2, t_2 | x_3, t_3) dx_2. \tag{2.10}$$

2.3 Formalism and Algorithms for Stochastic Dynamical System

Rate equations for the probabilities of the states are the stochastic analogue of the rate equations for the reagent concentration of the deterministic case. The rate equations of the probabilities are known as *master equations*. They are differential equations that describe the evolution of the probabilities for Markov processes for systems that jump from one to other state in a continuous time. Their use is widespread when the number of available states is discrete, as in the applications to chemical reactions or epidemic spreading. However, for chemical networks of realistic size and complexity yet their direct solutions have remained elusive [3, 4], and stochastic simulations algorithms offer a practical valuable alternative. In the next, we describe the two approaches, the one of master equations and that of stochastic simulation algorithms.

2.3.1 The Master Equation

The master equation is a differentiable form of Chapman–Kolmogorov equation. In some cases, the term *master equation* is used only for jump processes. Such processes are typical of discontinuous motion, for which there is a bounded and non-null transition probability per unit of time.

$$w(x|y, t) = \lim_{\Delta t \to 0} \frac{p_\to (x, t + \Delta t | y, t)}{\Delta t}$$

for some y such that $|x - y| > \varepsilon$. Here $w(x|y; t) = w(x|y)$. The master equation for jump processes can be written as

$$\frac{\partial p(x, t)}{\partial t} = \int \big(w(x|x')p(x', t) - w(x'|x)p(x, t) \big).dx' \qquad (2.11)$$

The master equation has a very intuitive interpretation. The first part of the integral is the *gain of probability* from the state x', and the second part is the *loss of probability* to x'. The solution is a probability distribution for the state space.

A reaction R is defined as a jump from the state $\mathbf{X}$ to a stare $\mathbf{X}_R$, where $\mathbf{X}, \ \mathbf{X}_R \in \mathbb{Z}_+^N$. The propensity $w(\mathbf{X}_R) = \tilde{v}(\mathbf{X})$ is the probability for transition from $\mathbf{X}_R$ to $\mathbf{X}$ per unit time. A reaction can be written as

$$\mathbf{X}_R \xrightarrow{w(\mathbf{X}_R)} \mathbf{X}.$$

The difference in molecules numbers $\mathbf{n}_R = \mathbf{X}_R - \mathbf{X}$ is used to write the master equation (2.11) for a system with M reactions

$$\frac{dp(\mathbf{X}, t)}{dt} = \sum_{i=1}^{M} w(\mathbf{X} + n)p(\mathbf{X} + \mathbf{n}_R, t) - \sum_{i=1}^{M} w(\mathbf{X})p(\mathbf{X}, t). \qquad (2.12)$$

This special case of master equations is called the *chemical master equation* (CME) [5, 6]. It is fairly easy to write: however, solving it is quite another matter. The number of problems for which the CME can be solved analytically is even fewer than the number of problems for which the deterministic reaction rate equations can be solved analytically. Attempts to use master equation to construct tractable time-evolution equations are also usually unsuccessful, unless all the reactions in the system are simple monomolecular reactions [7].

Master equations used to model chemical kinetics are usually referred to as *chemical master equation* (CME). Analytical solutions of the CME are possible to calculate only for simple special cases. CME suffers from the curse of dimensionality [8] as the size of the state space grows exponentially with the number of species involved. Due to this reason, CME is used to model the dynamics of gene regulatory networks [9–12] and protein–protein interaction networks, where the numbers of molecules of most species are around tens or (more rarely) around hundreds. In most applications to biological networks of realistic size, it is impossible to solve the CME directly, as the number of reachable states can be very large or infinite [13, 14].

Traditionally, researchers use the indirect approach of simulating trajectories by the stochastic simulation algorithm and variants such as τ-leaping [15]. Unfortunately, these approaches require a large number of trajectories when it comes to finding the probability distribution of the underlying Markov process. The finite state projection (FSP) is an alternative approach that seeks to approximate directly this distribution by restricting the CME to a more tractable size [16–23]. This method has recently been integrated into a larger framework that utilizes novel single-cell experimental methods to successfully identify predictive models of several gene regulatory networks [24], as well as to compare different CME models of single-cell, or single-molecule responses [25]. Unfortunately, more complex biological systems require a large state space that is beyond the reach of the standard FSP. This challenge motivated many developments, e.g. Krylov-FSP [26, 27], multiple-time-interval FSP [28], wavelet [29], quantized tensor train decomposition [30–32], SSA-driven FSP [18, 21] and many others.

Although these methods are promising for the purpose of solving the CME, the most widely used tool is the Gillespie stochastic simulation algorithm and its most current and most efficient variants [33, 34].

2.3.2 Stochastic Simulation Algorithm

In this section, we introduce the foundation of the stochastic simulation algorithm (SSA) of Gillespie. If we are given that the system is in the state $\mathbf{X} = \{X_1, \ldots, X_N\}$ at time t, computing its stochastic evolution means "moving the system forward in time". In order to do that, we need to answer two questions: (i) When will the next reaction occur? (ii) What kind of reaction will it be? Because of the essentially random nature of chemical interactions, these two questions are answerable only in a probabilistic way.

Let us introduce the function $P(\tau, \mu)$ defined as the probability that, given the state $\mathbf{X}$ at time t, the next reaction in the volume V will occur in the infinitesimal time interval $(t + \tau, t + \tau + d\tau)$ and will be an R_μ reaction. $P(\tau, \mu)$ is called *reaction probability density function*, because it is a joint probability density function on the space of the *continuous* variable τ $(0 \leq \tau < \infty)$ and the *discrete* variable μ $(\mu = 1, 2, \ldots, M)$.

The values of the variables τ and μ will give us answer to the two questions mentioned above. $P(\tau, \mu)$ can be written as the product of $P_0(\tau)$, the probability that given the state $\mathbf{X}$ at time t, no reaction will occur in the time interval $(t, t + dt)$, times $a_\mu d\tau$, the probability that an R_μ reaction will occur in the time interval $(t + \tau, t + \tau + d\tau)$

$$P(\mu, \tau)d\tau = P_0(\tau)a_\mu dt. \tag{2.13}$$

In turn, $P_0(\tau)$ is given by

$$P_0(\tau' + d\tau') = P_0(\tau')\left[1 - \sum_{i=1}^{M} a_i d\tau'\right],\tag{2.14}$$

where $[1 - \sum_{i=1}^{M} a_i d\tau']$ is the probability that no reaction will occur in time $d\tau'$ from the state $\mathbf{X}$. Therefore

$$P_0(\tau) = \exp\left[-\sum_{i=1}^{M} a_i \tau\right]\tag{2.15}$$

Inserting (2.14) into (2.13), we find the following expression for the reaction probability density function

$$P(\mu, \tau) = \begin{cases} a_\mu \exp(-a_0 \tau) & \text{if } 0 \le \tau < infty \\ 0 & otherwise \end{cases},\tag{2.16}$$

where a_μ is the reaction propensity for the reaction R_μ expressed as the product of the stochastic rate constant c_μ and the number h_μ of distinct combination of reactant molecules of R_μ

$$a_\mu = c_\mu \cdot h_\mu.\tag{2.17}$$

$$a_0 \equiv \sum_{i=1}^{M} a_i \equiv \sum_{i=1}^{M} h_i c_i.\tag{2.18}$$

The expression for $P(\mu, \tau)$ in (2.16) is, like the master equation in (2.12), a rigorous mathematical consequence of the fundamental hypothesis of stochastic chemical kinetics [7, 34, 35]. Notice finally that $P(\tau, \mu)$ depends on all the reaction constants (not just on c_μ) and on the current numbers of all reactant species (not just on the R_μ reactants).

On each step, the direct method of Gillespie algorithm generates two random numbers r_1 and r_2 from a set of uniformly distributed random numbers in the interval $(0, 1)$. The time for the next reaction to occur is given by $t + \tau$, where τ is given by

$$\tau = \frac{1}{a_0} \ln\left(\frac{1}{r_1}\right).\tag{2.19}$$

The index μ of the occurring reaction is given by the smallest integer satisfying

$$\sum_{j=1}^{\mu} a_j > r_2 a_0.\tag{2.20}$$

The system states are updated by $X(t + \tau) = X(t) + v_\mu$, and then the simulation proceeds to the next occurring time.

We now summarize the steps of direct method of Gillespie simulation algorithm.

1. Initialization: Set the initial numbers of molecules for each chemical species; input the desired values for the M reaction constants $c_1, c_2, \ldots, c_M$. Set the simulation time variable t to zero and the duration T of the simulation.
2. Calculate and store the propensity functions a_i for all the reaction channels ($i = 1, \ldots, M$), and a_0.
3. Generate two random numbers r_1 and r_2 in $Unif(0, 1)$.
4. Calculate τ according to (2.19).
5. Search for μ as the smallest integer satisfying (2.20).
6. Update the states of the species to reflect the execution of μ (e.g. if R_μ: $S_1 + S_2 \to 2S_1$, and there are X_1 molecules of the species S_1 and X_2 molecules of the species S_2, then increase X_1 by 1 and decrease X_2 by 1). Set $t \leftarrow t + \tau$.
7. If $t < T$ then go to step 2, otherwise terminate.

Note that the random pair (τ, μ), where τ is given by (2.19) and μ by (2.20), is generated according to the probability density function in (2.16). A rigorous proof of this fact may be found in [35]. Suffice here to say that (2.19) generates a random number τ according to the probability density function

$$P_1(\tau) = a_0 \exp(-a_0\tau), \tag{2.21}$$

while (2.20) generates an integer μ according to the probability density function

$$P_2(\mu) = \frac{a_\mu}{a_0} \tag{2.22}$$

and the stated result follows because

$$P(\tau, \mu) = P_1(\tau) \cdot P_2(\mu).$$

To generate random numbers between 0 and 1, we can do as follows. Let $F_X(x)$ be a distribution function of an exponentially distributed variable X and let $U \sim Unif[0, 1)$ denote an uniformly distributed random variable U on the interval $[0, 1)$.

$$F_X(x) = \begin{cases} 1 - e^{-ax} & \text{if } x \geq 0 \\ 0 & \text{if } x < 0 \end{cases}. \tag{2.23}$$

$F_X(x)$ is a continuous non-decreasing function, and this implies that it has an inverse F_X^{-1}. Now, let $X(U) = F_X^{-1}(U)$ and we get the following:

$$P(X(U) \leq x) = P(F_X^{-1}(U) \leq x) = P(U \leq F_X(x)) = F_X(x). \tag{2.24}$$

It follows that

$$F_X^{-1}(U) = -\frac{\ln(1-U)}{a} \sim Exp(a). \tag{2.25}$$

In returning to step 1 from step 7, it is necessary to re-calculate only those quantities a_i, corresponding to the reactions R_i whose reactant population levels were altered in step 6; also a_0 must be re-calculated simply by adding to it the difference between each newly changed a_i value and its corresponding old value.

Since this algorithm uses M random numbers per iteration, it takes time proportional to M to update the a_i's. This results in a computational complexity that scales linearly as the problem size increases. The original Gillespie algorithm is therefore inefficient for large problems, which has prompted the development of several alternative formulations with improved scaling properties. A comprehensive review of the literature about efficient alternatives to Gillespie original algorithm can be found in [34, 36–39]. The majority of the current alternatives are approximation of variables' estimates defined by the Gillespie original methods, or hybrid deterministic/stochastic algorithms.

2.4   Stiff Dynamical Systems and Hybrid Stochastic Algorithms

In the SSA, however, every reaction event has to be resolved explicitly such that it becomes numerically inefficient when the system's dynamics includes fast reaction processes or species with high population levels [1, 40].

The co-existence of fast and slow reactions makes the dynamic system *stiff*. Stiff differential equations arise in equations due to the existence of greatly differing time constants. Indeed, in stiff dynamical systems, the largest fluctuating species require the most time to simulate stochastically because exact stochastic simulation algorithms scale with the number of reaction events [41].

Stiffness makes hard the numerical integration of the differential equations and their calibration [42–45]. As an example of stiff dynamics, we consider the following system of reactions:

$$A \underset{k_2}{\overset{k_3}{\rightleftharpoons}} B, \quad A \overset{k_1}{\rightarrow} \emptyset, \quad B \overset{k_4}{\rightarrow} \emptyset \tag{2.26}$$

where k_1, k_2, k_3 and k_4 are real positive numbers. Assuming that the dynamic of this system follows the law of mass action, the differential equations for the concentrations of A and B are

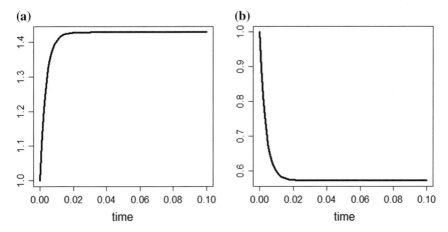

Fig. 2.1 Numerical solutions of the ordinary differential equation system (2.27)

$$\frac{d[A]}{dt} = -k_1[A] + k_2[B]$$

(2.27)

$$\frac{d[B]}{dt} = k_3[A] - k_4[B].$$

If $k_1 = k_3 = 80$ and $k_2 = k_4 = 200$ (in units of time^{-1}), and $[A](0) == [B](0) = [1]$, the system is stiff. Its simulation in the time interval $[0, 1]$ is shown in Fig. 2.1. In spite of the rapid decrease of B which is the only source from which A feeds, after a rapid initial growth, A takes a great deal of time to decrease

Hybrid stochastic–deterministic models may provide a conceptual and computational basis for all these observations, see, for example, [1, 46], and also the recent literature in [13, 47–54]. Commonly, in hybrid algorithms, fast reactions are approximated as continuous-deterministic processes or replaced by quasi-stationary distributions either in a stochastic or deterministic context, while slow reaction involving a low number of molecules is modelled with discrete-stochastic formalisms. The main theoretical concern about these approaches relates to the classification of the reactions into slow and fast processes. As also in a previous study of the author [13], it is not always possible to make a clear-cut distinction. The choice of the threshold on concentration and/or velocity of reaction is context-specific, and often left to the a priori knowledge of the user. Sometimes, in the absence of a priori knowledge, the threshold is set according to a thumb rule dictated by the common sense, for instance, it is decided to treat as stochastic the dynamics of species whose number of molecules is about or less than few tens, and as deterministic the dynamics of species whose amount counts hundreds or thousands (and above) of molecules. Furthermore, in [13], the author shows the choice of an a priori classification of deterministic and stochastic reactions in two sets whose elements (reactions) always remain the same during all the simulation times, which can lead to unreliable results. Lecca et al.

[13] highlighted the need to take into account the possibility that reaction initially classified as stochastic can turn to deterministic and vice versa, as the concentration of the species changes and may exceed or fall below the set threshold. It is also worth today that to overlook the possibility of transitions from a stochastic dynamics to a deterministic results in an increment of the program execution time.

2.5 Exercises

Exercise 2.1 We describe an enzyme-catalysed substrate conversion by the three reactions R_1: $E + S \longrightarrow ES$, R_2: $ES \longrightarrow E + S$, R_3: $ES \longrightarrow E + P$. This network involves four chemical species, namely, enzyme (E), substrate (S), complex (ES) and product (P) molecules. Let k_1, k_{-1} and k_2 be the rate constants of the reactions R_1, R_2 and R_3, respectively. Assuming that the intermediate complex ES remains approximately equal to the total amount of enzyme $E_T = E(t) + ES(t)$,

$$E_T \approx \frac{S}{(Km + S)}.$$

Prove that

$$\frac{dP}{dt} = \frac{V_{\max} S}{K_M + S},$$

where $V_t extrm{max} = k_2 E_T$ and $K_M = \frac{k_{-1}+k_2}{k_1}$. $V_t extrm{max}$ is the maximum rate of product formation under substrate saturation (i.e. $S \gg K_M$) and K_M is the substrate concentration at which the product formation rate is $V_{\max}/2$.

As reported by Sanft et al. [55], the Michaelis–Menten approximation can be viewed as a model in which we eliminate the species E and ES and replace the full system with the reduced model

$$S \xrightarrow{c_M} P,$$

where

$$c_M = \frac{V_{\max}}{K_M + S}.$$

The rate constants k_1, k_{-1} and k_2 often are not accessible to a direct measurement, whereas K_M and $V_{\max}$ can be directly measured. In this situation, would it be possible to simulate the kinetics of Michaelis–Menten with a stochastic simulation algorithm like that of Gillespie? If so, explain how.

Exercise 2.2 Consider the system of differential equations

$$\frac{dy_1}{dt} = -k_1 y_1 + k_2 y_2$$

$$\frac{dy_2}{dt} = k_3 y_1 - k_4 y_2,$$

where $k_1 = k_3 = 0.01$ and $k_2 = k_4 = 100$. Answer the following questions:

1. Can we say that this system is stiff? If so, explain why.
2. Calculate the numerical solutions of this system.
3. Draw the graph representing this system (y_1 and y_2 are the nodes of the graph).
4. Using the graph plotted in the previous point, simulate the system with the Gillespie algorithm.
5. Compare the computational time employed to find the deterministic solution, with the one employed to terminate the stochastic simulation. Which is the greatest?

Exercise 2.3 We consider a gene expression model, which involves two chemical species, namely, mRNA (M) and protein (P). Transcription of a gene into mRNA is modelled by reaction $R_1 : \emptyset \xrightarrow{k_1} M$, translation of mRNA into protein by $R_2 : M \xrightarrow{k_2} M + P$, degradation of mRNA by $R_3 : M \xrightarrow{k_3} \emptyset$ and degradation of proteins by $R_4 : P \xrightarrow{k_4} \emptyset$.

1. Write the state-change matrix of this system, i.e. a matrix whose row names are the names of species and column names are the reactions. The entries of this matrix are the number of molecules defined by the stoichiometry assumed for this system.
2. Write the vector of reaction propensities.
3. Estimate the complexity of the direct Gillespie algorithm for this case study.
4. Explain why this system has a hybrid stochastic/deterministic dynamics.

Exercise 2.4 Write the chemical master equation for the following system:

$$A + B \xrightarrow{k_1} C \xrightarrow{k_2} \emptyset.$$

knowing that probability per unit of time that A binds to B is p_1 and the probability that C degrades is p_2.

References

1. Lecca P, Laurenzi I, Jordan F. Deterministic versus stochastic modelling in biochemistry and systems biology. Woodhead Publishing Series in Biomedicine No. 21; 2012.
2. Lecca P. Stochastic chemical kinetics: a review of the modelling and simulation approaches. Biophys Rev. 2013;5(4):323–45.
3. Kazeev V, Khammash M, Nip M, Schwab C. Direct solution of the chemical master equation using quantized tensor trains. PLoS Comput Biol. 2014;10(3):e1003359 March.
4. Paola L, Angela R. Theoretical physics for biological systems. Boca Raton: CRC Press; 2019.
5. van Kampfen NG. Stochastic processes in physics and chemistry. Amsterdam: Elsevier; 1992.
6. McQuarrie DA. Stochastic approach to chemical kinetics. J Appl Prob. 1967;4:413–78.

7. Gillespie DT. Exact stochastic simulation of coupled chemical reactions. J Phys Chem. 1977;81(25).
8. Kazeev V, Khammash M, Nip M, Schwab C. Direct solution of the chemical master equation using quantized tensor trains. PLoS Comput Biol. 2014;10(3):e1003359 Mar.
9. Jaroslav A. A hybrid of the chemical master equation and the gillespie algorithm for efficient stochastic simulations of sub-networks. PLoS ONE. 2016;11:1–22.
10. Calder M, Gilmore S, Khanin R, Higham DJ. Computational methods in systems biology chemical master equation and Langevin regimes for a gene transcription model. Theor Comput Sci. 2008;408(1):31–40.
11. Hemberg M, Barahona M. Perfect sampling of the master equation for gene regulatory networks. Biophys J. 2007;93(2):401–10.
12. Booth HS, Burden CJ, Hegland M, Santoso L. A stochastic model of gene regulation using the chemical master equation. In: Fagerberg J, Mowery DC, Nelson RR, editors. Mathematical modeling of biological systems, vol. I. Birkhäuser: Boston; Springer: Berlin; 2007. p. 71–81.
13. Lecca P, Bagagiolo F, Scarpa M. Hybrid deterministic/stochastic simulation of complex biochemical systems. Mol BioSyst. 2017;13:2672–86.
14. Kaznessis YN, Smadbeck P. Solution of chemical master equations for nonlinear stochastic reaction networks. Curr Opin Chem Eng. 2014;5:90–5.
15. Cao Y, Gillespie DT, Petzold LR. Efficient step size selection for the tau-leaping simulation method. J Chem Phys (AIP Publishing). 2006;124(4):044109. https://doi.org/10.1063/1.2159468.
16. Gupta A, Mikelson J, Khammash M. A finite state projection algorithm for the stationary solution of the chemical master equation. J Chem Phys. 2017;147(15):154101.
17. Munsky B, Khammash M. A finite state projection algorithm for the stationary solution of the chemical master equation. J Chem Phys. 2017;147(15):154101.
18. Dinh KN, Sidje RB. Understanding the finite state projection and related methods for solving the chemical master equation. Phys. Biol. 2016;13(3):035003.
19. Hjartarson A, Ruess J, Lygeros J. Approximating the solution of the chemical master equation by combining finite state projection and stochastic simulation. In: 2013 IEEE 52nd annual conference on decision and control (CDC); 2013.
20. Sunkara V, Hegland M. An optimal finite state projection method. Proc Comput Sci. 2010;1(1):1579–86 ICCS 2010.
21. MacNamara S, Sidje R, Burrage K. An improved dynamic finite state projection algorithm for the numerical solution of the chemical master equation with applications. ANZIAM J. 2007;48:413–35.
22. MacNamara S, Burrage K, Sidje RB. Multiscale modeling of chemical kinetics via the master equation. Multiscale Model Simul. 2008;6(4):1146–68.
23. Munsky B, Khammash M. The finite state projection algorithm for the solution of the chemical master equation. J Chem Phys. 2006;124(4):044104.
24. Weber L, Raymond W, Munsky B. Identification of gene regulation models from single-cell data. bioRxiv; 2017.
25. Fox Z, Neuert G, Munsky B. Finite state projection based bounds to compare chemical master equation models using single-cell data. J Chem Phys. 2016;145(7):074101.
26. Vo H, Sidje R. Improved Krylov-FSP method for solving the chemical master equation. In: World congress on engineering and computer science. Lecture notes in engineering and computer science, vol 2226; 2016, p. 521–6.
27. Dinh KN, Sidje RB. An application of the Krylov-FSP-SSA method to parameter fitting with maximum likelihood. Phys Biol. 2017;14(6):065001.
28. Munsky B, Khammash M. A multiple time interval finite state projection algorithm for the solution to the chemical master equation. J Comput Phys. 2007;226(1):818–35.
29. Udrescu T, Jahnke T. An adaptive method for solving chemical master equations using a sparse wavelet basis. In: AIP conference proceedings, vol. 489, 04; 2009, p. 1168.
30. Kazeev V, Khammash M, Nip M, Schwab C. Direct solution of the chemical master equation using quantized tensor trains. PLoS Comput Biol. 2014;10(3):e1003359.

31. Vo H, Sidje R. Solving the chemical master equation with the finite state projection and inexact uniformization in quantized tensor train format. In: 5th international conference on computational and mathematical biomedical engineering - CMBE2017; 2017, p. 1108.
32. Vo H, Sidje R. An adaptive solution to the chemical master equation using tensors. 2017;147:044102.
33. Gillespie TD. Gillespie algorithm for biochemical reaction simulation. New York: Springer; 2013.
34. Gillespie DT. Stochastic simulation of chemical kinetics. Annu Rev Phys Chem. 2007;58(1):35–55.
35. Gillespie DT. A general method for numerically simulating the stochastic time evolution of coupled chemical species. J Comput Phys. 1976;22:403–34.
36. Albert J. A hybrid of the chemical master equation and the gillespie algorithm for efficient stochastic simulations of sub-networks. PLoS ONE. 2016;11:03.
37. Sanft KR, Othmer HG. Constant-complexity stochastic simulation algorithm with optimal binning. J Chem Phys. 2015;143(7):074108.
38. Wilkinson DJ. Stochastic modelling for systems biology. Boca Raton: CRC Press, Taylor & Francis Group; 2012.
39. Jack J, Păun A, Rodríguez-Patón A. A review of the nondeterministic waiting time algorithm. 2011;10(1):139–49.
40. Resat H, Petzold L, Pettigrew MF. Kinetic modeling of biological systems. In: Methods in molecular biology. Humana Press: Totowa, 2009, p. 311–35.
41. Haseltine EL, Rawlings JB. Approximate simulation of coupled fast and slow reactions for stochastic chemical kinetics. J Chem Phys. 2002;117:6959–69.
42. Zhang S, Li J. Explicit numerical methods for solving stiff dynamical systems. J Comput Nonlinear Dyn. 2011;6(4):041008.
43. Matsubara Y, Kikuchi S, Sugimoto M, Tomita M. BMC Bioinform. 2006;7(1):230.
44. Rué P, Villà-Freixa J, Burrage K. Simulation methods with extended stability for stiff biochemical kinetics. BMC Syst Biol. 2010;4(1):110.
45. Petzold L. Automatic selection of methods for solving stiff and nonstiff systems of ordinary differential equations. SIAM J Sci Stat Comput. 1983;4(1):136–48 March.
46. Wilkinson D. Stochastic modelling for quantitative description of heterogeneous biological systems. Nat Rev Genet. 2009;10:122–33.
47. Menz S, Latorre JC, Schütte C, Huisinga W. Hybrid stochastic-deterministic solution of the chemical master equation. SIAM Interdiscip J Multiscale Model Simul (MMS). 2012;10:1232–62.
48. Lachor P, Puszynski K, Polanski AJ. Deterministic models and stochastic simulations in multiple reaction models in systems biology. J Biotechnol, Comput Biol Bionanotechnol. 2011;92(3):265–80.
49. Perkins TJ, Wilds R, Glass L. Robust dynamics in minimal hybrid models of genetic networks. Philos Trans A Math Phys Eng Sci. 2010;368(1930):4961–75.
50. Crudu ADA, Radulescu O. Hybrid stochastic simplifications for multiscale gene networks. BMC Syst Biol. 2009;3(1):1–25.
51. Pahle J. Biochemical simulations: stochastic, approximate stochastic and hybrid approaches. Brief Bioinform. 2009;10(1):53–64.
52. Samant A, Ogunnaike BA, Vlachos DG. A hybrid multiscale Monte Carlo algorithm (HyMSMC) to cope with disparity in time scales and species populations in intracellular networks. BMC Bioinform. 2007;8(1):1–23.
53. Salis H, Sotiropoulos V, Kaznessis YN. Multiscale Hy3S: hybrid stochastic simulation for supercomputers. BMC Bioinform. 2006;7(1):1–21.
54. Alur R, Belta C, Ivancic F, Kumar V, Mintz M, Pappas GJ, Rubin H, Schug J. Hybrid modeling and simulation of biomolecular networks. Lecture notes in computer science, 2034:19–32, 2001.
55. Sanft KR, Gillespie DT, Petzold LR. Legitimacy of the stochastic michaelis–menten approximation. IET Syst Biol. 2011;5(1):58–69 January.

Chapter 3
Model Identifiability

Abstract Once we have built a model to describe the dynamics of a network, in order to simulate this dynamic, that is, the evolution in the time of the network, we need to know the parameters of the model. Very often the values of the kinetic constants in a network of biochemical interactions, or more generally the arcs' weights on the network define the force and direction of the interaction between nodes, are obtained from experimental data through various regression and inference techniques. In this chapter, we will tackle a problem that is upstream of the parameter estimation, that is, the possibility to infer them from the data. The problem is known as *identifiability*. Identifiability is a fundamental prerequisite for model identification. It concerns uniqueness of the model parameters determined from experimental observations. This paper specifically deals with structural or a priori identifiability: whether or not parameters can be identified from a given model structure and experimental measurements. Since experimental data are usually affected by uncertainties, this question is known as *practical identifiability*. Non-identifiability of parameters induces non-observability of trajectories, reducing the predictive power of the model. We will discuss here a method of parameter identifiability based on the observability rank test and how much it is suitable to handle noisy observations.

3.1 Parameter Identifiability: The State of the Art of the Problem

Quantitative modelling of biochemical networks is the basis for the analysis of the dynamics of biological processes. However, some parameters in the models are only roughly known and many others are unknown, so they must be estimated.

The recent literature reports many examples of methods for parameter estimation both in deterministic and stochastic models. Here, we briefly mention the most recent ones and refer the reader to an extensive review of the state of the art reported by Chou and Voit [1]. Goel et al. in [2] and Voit et al. in [3] discussed the problem of lack of convergence of search algorithms on generalized mass action law and proposed a novel methodological framework called dynamic flux estimation for estimating parameters for models of metabolic systems from time series data. Polisetty et al. [4]

© The Author(s), under exclusive license to Springer Nature Switzerland AG 2020
P. Lecca, *Identifiability and Regression Analysis of Biological Systems Models*,
SpringerBriefs in Statistics, https://doi.org/10.1007/978-3-030-41255-5_3

suggested global optimization techniques as alternative to traditional local methods. Rodrigez-Fernandez et al. [5] developed a hybrid stochastic–deterministic global optimization method. Moles et al. [6] explored several state-of-the-art deterministic and stochastic global optimization techniques and compared their accuracy and effectiveness on non-linear biochemical dynamic models. Tian et al. [7] presented a simulated maximum likelihood method to evaluate parameters in stochastic models described by stochastic differential equations. They proposed different types of transitional probability and a genetic optimization algorithm to search for optimal reaction rates. Chou et al. [8] developed an alternate regression method which dissects the parameter inference problem into iterative steps of linear regression. Sugimoto et al. [9] developed a computational technique based on genetic programming that simultaneously generates biochemical equations and their parameters from time series data. Reinker et al. [10] proposed the approximate maximum likelihood method and the singular value decomposition likelihood method that estimate stochastic reaction constants from molecule count data measured with errors at discrete time points. Tools for parameter fitting through regression or maximum likelihood methods can be found as integral part of simulation tools (e.g. Copasi [11]), but there exist also stand alone, like PET [12] and BioBayes [13]. BioBayes has been recently developed by Vyshemirsky and Girolami [14, 15]: it is a software package which provides a framework for Bayesian parameter estimation and model ranking over models of biochemical systems defined using ordinary differential equations. Boys [16], Golightly [17] and Wilkinson [18, 19] developed Bayesian model-based inference techniques for discrete models. Finally, the authors themselves proposed a new maximum likelihood method and a software tool to infer the values of the rate constants of a system of chemical reaction from experimental time series of reactant concentrations [20].

Along with the huge amount of efforts in developing methods able to parametrize non-linear stiff models of biochemical dynamics, more and more attention has been focused on the question of whether or not parameters can in fact be determined uniquely from a given model and choice of measurements. Namely, a crucial problem of parameter identification is to be able, before the data have been analysed, to establish if all the unknown parameters of a model can be uniquely recovered from data. This is the question of identifiability that is focused on here. Determining the identifiability of a model of biochemical network serves several purposes. Since model parameters have physical significance, it is important to know if their values can be determined from the observed data and the given model. The most of the numerical search procedure for parameter estimation may be not effective for problems in which a unique solution does not exist, and therefore they offer no guarantee that available measurements will yield physically meaningful parameter values.

We focus on identifiability of non-linear systems in continuous time as this is generally the result of the use of the generalized mass action law for modelling biochemical reaction networks. Several methods are available for checking identifiability of this type of system such as those found in [21–28]. However, there is no well-established identifiability method for solving the identifiability of biochemical continuous-time non-linear dynamical systems. Many of these methods restate the question of identifiability as one of the observabilities [29]. To determine the observability (and the controllability, the dual of observability [30]) of a system means to establish if there exist relations linking the state variables to the inputs, outputs and their time derivatives and thus locally defining them uniquely in terms of measurable quantities without the need for knowing the initial conditions [22]. The concept of observability and the methods to prove the observability property of a mathematical model of a physical/chemical phenomenon belong to the methodologies of systems theoretic analysis. The common denominator of these methodologies is the assumption that the dynamic model describing the internal structure is an input–output map. In this system, we may have no knowledge of the initial conditions for the state variables (or of the parameter values), but we assume to have measurements of the outputs good enough to allow us to model them as analytic functions of time and to know with a good precision the value of their time derivatives at time zero. Such a situation is not too unrealistic in molecular biology. The current experimental investigation of the cellular metabolism is just designed as an input–output experiment. Consider a culture of cells grown in a culture medium and the chemical reactions that take place in their metabolism. At the moment of writing, the ability to observe what occurs inside a single cell as far as metabolic fluxes are concerned is very limited. Therefore, in preparation for a mathematical description of the system, the cell is considered as a box where we see what goes in (for instant, the nutrients present in the reactor) and the secreted products, but we cannot observe/measure what happens inside. If we assume such a representation of the system and there is no relation binding the state variables to the inputs, outputs and their rate of change in time, the initial state of the system cannot be inferred from observing its input–output behaviour. In the biological setting we described above, the inability to deduce the initial conditions causes the impossibility to estimate a unique model parameter set from the experimental measurements. The lack of uniqueness of the solution with respect to the parameters means that there could be infinitely many parameter sets that reproduce exactly the same output for every input and thus that model parameters cannot be estimated from *any* experimental measurements.

3.2 Observability, Distinguishability and Identifiability

In this section, we report the definition of three important properties of a dynamical systems such as observability, distinguishability and identifiability, and the relationships among them. We summarize the work of Anguelova [31] of which we report and comment some definitions and explanations. We refer the reader to [31] for a more comprehensive exposition. For the comfortableness of the reader, we adopt the same notation of Anguelova [31]. Other useful and recent references on these topics can be found in [28, 32, 32–34] as evidence of the thriving research activity on these issues.

A non-linear dynamic system can be written in the following form:

$$\Sigma = \begin{cases} \frac{dx(t)}{dt} = f(x(t), u(t), \theta) = h_0(x) + h(x)u, \ x(0) = x_0 \\ y(t) = g(x(t), u(t), \theta) \end{cases},$$

where $x \in M \subseteq \mathbb{R}^n$ is the state vector, $u \in \mathbb{R}^k$ is the vector of inputs (i.e. control variables), $y \in \mathbb{R}^p$ is the vector of outputs (i.e. the observations) and $\theta \in \mathbb{R}^q$ is the vector of unknown parameters; h_0 and the k columns of h (denoted by h^i for $i = 1, \ldots, k$) are analytic vector fields defined on M. We assume as in [22] that the system is complete, i.e. for every bounded measurable input $u(t)$ and every $x_0 \in M$, there exists a solution to

$$\frac{dx}{dt} = f(x(t), u(t), \theta) \tag{3.1}$$

such that $x(0) = x_0$ and $x(t) \in M$, for all $t \in \mathbb{R}$. Following the notation used in [22], let U denote an open subset of M. Under these conditions, we introduce the following definition of *distinguishability*.

Definition 3.1 (U-distinguishability) A pair of points x_0 and x_1 in M are called *U-distinguishable* if there exists a measurable bounded input $u(t)$ defined on the interval $[0, T]$ that generates solutions $x_0(t)$ and $x_1(t)$ of Eq. (3.1), satisfying $x_i(0) = x_i$ such that $x_i(t) \in U$ for all $t \in [0, T]$ and $g(x_0(t)) \neq g(x_1(t))$ for some $t \in [0, T]$.

Let us denote by $J(x_0, U)$ the set of the points $x_1 \in U$ that are not U-distinguishable from x_0. The definition of U-distinguishability leads to the definition of *observability* and *local observability*.

Definition 3.2 (Observability) The system S is *observable* at $x_0 \in M$ if $I(x_0, M) = x_0$.

Although a system may be observable according to the above definition, it is still possible that there is an arbitrarily large interval of time in which two points of M cannot be distinguished from each other. A concept of local observability is then introduced to guarantee that to distinguish between the points of an open subset U of M, we do not have to go outside U, which sets also a limit to the time interval.

Definition 3.3 (Local observability) The system S is *locally observable* at $x_0 \in M$ if for every open neighbourhood U of x_0, $I(x_0, U) = x_0$.

Definition 3.4 (Distinguishability) The system S has the *distinguishability* property at $x_0 \in M$ if x_0 has an open neighbourhood V such that $J(x_0, M) \cap V = x_0$.

As for the observability, also for distinguishability a concept of locality needs to be introduced. In fact, although in a system having the distinguishability property, any point x_0 can be distinguished from neighbouring points, there could be arbitrarily large intervals of time in which the points are indistinguishable. In order to establish a limit on the time interval, the definition of *local distinguishability* is introduced.

Definition 3.5 (Local distinguishability) The system S has the *local distinguishability* property at $X - 0 \in M$ if x_0 has an open neighbourhood V such that for every open neighbourhood U of x_0, $I(x_0, U) \cap V = x_0$.

Local observability implies local distinguishability as we can set V equal to M. Therefore, if a system does not have the local distinguishability property at some x_0, it is not locally observable at that point either. Anguelova [31] note that it is the property of local distinguishability that lends itself to a test. This test is known as test of *observability rank*. Therefore, to prove the observability of a system, it is sufficient to prove that it holds the property of local distinguishability.

To determine if a system is local distinguishable, we consider the Lie differentiation of a C^∞ function ϕ on M by a vector field v on M, defined as follows:

$$L_v(\phi)(x) = \langle d\phi, v \rangle, \tag{3.2}$$

where $\langle \cdot \rangle$ denotes the scalar product and $d\phi$ the gradient of ϕ. The flow $\Phi(t, x)$ of a vector field v on M is by definition the solution of

$$\begin{cases} \frac{\partial}{\partial t}\Phi(t, x) = v(\Phi(t, x)) \\ \Phi(0, x) = x \end{cases}.$$

Observe that we have

$$L_v(\phi)(x) = \frac{d}{dt}\left(\phi(\Phi(t, x))\right)\Big|_{t=0}, \tag{3.3}$$

where $\phi(\Phi(t, x))$ can be expanded in the Lie series as follows:

$$\phi(\Phi(t, x)) = \sum_{l=0}^{\infty} \frac{t^l}{l!} L_v^l(\phi)(x). \tag{3.4}$$

For a constant input u, $f(x, u)$ defines a vector field on M and we can define the flow $\Phi(t, x)$ and the Lie series expansion of $g_i(\Phi(t, x))$ for $i = 1, \ldots, p$. Note that if two points x_0 and $x - 1$ in M are distinguishable by a bounded measurable input, then they must be distinguishable by a piecewise constant input due to the uniform convergence since the outputs depend continuously on the inputs. To prove this, consider the input such that

$$\begin{cases} u_i(t) = u_i^1, \ t \in [0, t_1) \\ u_i(t) = u_i^l, \ t \in [t_{l-1}, t_l), \quad l \geq 2 \end{cases}$$

where $i = 1, \ldots, p$ and $u_i^l \in \mathbb{R}$, and define the vector fields

$$\theta_l = h_0 + h u^l \tag{3.5}$$

and denote their corresponding flows by Φ_t^l. Therefore, the state reached at time t_l starting from x^0 at $t = 0$ can be expressed as the composition of the flow functions:

$$x(t_l) = \Phi_{t_l}^l \circ \ldots \circ \Phi_{t_1}^1(x_0). \tag{3.6}$$

The corresponding outputs become

$$y_i(t_l) = g_i(\Phi_{t_l}^l \circ \ldots \circ \Phi_{t_1}^1(x_0)). \tag{3.7}$$

The time derivative at zero of the output g_i can then be calculated and we can define

$$L_f g_i = L_{\theta_1} \ldots L_{\theta_l} g_i. \tag{3.8}$$

If two neighbouring points x_0 and x_1 are U-distinguishable instantaneously (which is the requirement for local distinguishability), then there exists a piecewise constant input u such that the sets of Lie series coefficients of $g_i(\Phi(ty, x_0))$ and $g_i(\Phi(t, x_1))$ differ for some i.

Consider now the linear map from M to the space $\mathcal{G}$ spanned by functions $L_f^k(g_i)$ at x_0 for $k \geq 0$ ($i = 1, \ldots, p$) for all vector fields $f(x, u)$ defined by piecewise constant inputs u:

$$\mathcal{G} = \{L_{h^{i_1}} L_{h^{i_2}} \ldots L_{h^{i_r}}(g_i) : r \geq 0, i_j = 0, \ldots, k, \ i = 1, \ldots, p\}. \tag{3.9}$$

Thus, the space $d\mathcal{G}$ spanned by the gradients of the elements of $\mathcal{G}$ is

$$d\mathcal{G} = \{d\phi : \phi \in \mathcal{G}\}. \tag{3.10}$$

The dimension of $\mathcal{G}$ determines the local distinguishability property. For each $x \in M$, let $d\mathcal{G}(x)$ be the subspace of the cotangent space at x obtained by evaluating the elements of $d\mathcal{G}$ at x. The rank of $d\mathcal{G}(x)$ is constant in M except at certain singular points, where the rank is smaller because the system is analytic [22, 35]. Then

$$\dim d\mathcal{G} = \max_{x \in M} \left(\dim d\mathcal{G}(x)\right). \tag{3.11}$$

Theorem 3.1 *The system S has the local distinguishability property for all x in an open dense set of M if and only if*

$$\dim d\mathcal{G} = n. \tag{3.12}$$

Definition 3.6 (A priori identifiability) A given parameter θ_i is a priori or *structurally* identifiable if there exist a unique solution to Σ for θ_i. A parameter with a countable or uncountable number of solutions is considered locally identifiable or unidentifiable, respectively.

3.3 Observability Rank Test

Referring back to what was said in the previous section, a system is said to be observable if any initial state x_0 can be estimated from the control $u(t)$ and the measurements $y(t)$.

Consider a continuous-time non-linear system described by

$$\begin{cases} \dot{x} = f(x) \\ y = h(x) \end{cases}, \tag{3.13}$$

where, in general, the state $x \in \mathbb{C}^n$ and the output $y \in \mathbb{C}$. The elements f_i ($i = 1, 2, \ldots, n$) and h are complex meromorphic functions of x.

For a linear system, the observability Popov–Belevitch–Hautus (PBH) test is one of the criteria for determining observability [36]. This criterion states the following. Suppose that $f = Ax$ and $h = c^T x$ in (3.13), where $A \in \mathbb{C}^n$, n and $c \in \mathbb{C}^n$. System (3.13) is observable if and only if

$$\text{rank} \begin{pmatrix} \lambda I - A \\ c^T \end{pmatrix} = n, \quad \forall \lambda \in \mathbb{C}^n. \tag{3.14}$$

Kawano et al. [36] generalized condition (3.14) to non-linear systems, by exploring the relationship between observability and the following condition:

$$\text{rank} \begin{pmatrix} \lambda I - \frac{\partial f(x)}{\partial x} \\ \frac{\partial h(x)}{\partial x} \end{pmatrix} = n, \quad \forall \lambda \in \mathbb{K}. \tag{3.15}$$

Differently from that considered in condition (3.14), the field considered in condition (3.15) is the field of meromorphic functions, and it is not required that condition (3.15) holds for all $x \in \mathbb{C}^n$.

Kawano et al. [36] introduced a pseudo-linear transformation $\delta : \daleth \longrightarrow \mathbb{K}$ and came to formulate the following theorem.

Theorem 3.2 *The system (3.14) is observable if the condition*

$$\text{rank} \begin{pmatrix} \lambda I - \left[\delta(J_\phi) + J_\phi \frac{\partial f(x)}{\partial x} \right] J_\phi^{-1} \\ \frac{\partial h(x)}{\partial x} J_\phi^{-1} \end{pmatrix} v \neq 0 \tag{3.16}$$

holds for all $\phi \in \text{Diff}_\mathbb{K}^n(\mathbb{C})$, $\lambda \in \mathbb{K}$, and $v \in \mathbb{K}^n \setminus \{0\}$. $\phi(x) \in \mathbb{K}^n$, J_ϕ is the Jacobian matrix of ϕ, $\text{Diff}_\mathbb{C}^n \subset \mathbb{K}^n$ is the set of $\phi \in \mathbb{K}^n$ such that $\text{rank}_\mathbb{K} J_\phi = n$. The coordinate transformation ϕ transforms a system into its feed-forward form.

The feed-forward form plays important role in the observability theory of non-linear systems. Once a system is in the feed-forward form, we can find a solution to the system equations simply by integrating recursively the respective state equations [37].

3.4 Biochemical Networks and Identifiability

To give a concrete case of study on model identifiability, we consider a metabolic network with M metabolites and N chemical reactions. Kinetic models of metabolism can be described by the following equation [27, 31, 38]:

$$\dot{m}_j = u_j + \sum_i v_{ij} \cdot r_i(c, p_i) \qquad (3.17)$$

for each metabolite j, and $i = 1, 2, \ldots, M, j = 1, 2, \ldots, N$. The coefficients v_{ij} are the stoichiometric coefficients associated to each reaction, and r_i the flux associated to reaction i, that is,[1]

$$\dot{r}_i = k_i \prod_{j=1}^{M} c_i^{v_{ij}}.$$

Assuming that the outputs $y(t)$ are the concentrations c of a set of species, observability aims to identify a minimum set of species from whose concentrations it is possible to determine the concentration of all other species [27].

Liu et al. [27] proposed an interesting network-based approach to determine observability of complex biochemical networks, and introduced the concept of *target observability*. Liu et al. state

"Notwithstanding the fundamental importance of full observability, aiming to derive the state of each variable in a system, for most applications it is sufficient to infer the state of a certain subset of variables, that we call target variables, like the concentrations of metabolites whose activities are altered by a disease. If those target variables cannot be directly measured, we can invoke target observability, identifying the optimal sensor(s) that can infer the state of the target variables, thus discovering the optimal biomarkers for the respective disease".

Liu et al. used a genetic algorithm to select optimal sensors, according to the following principles:

1. The state of a target node x_T can be observed from a sensor node $x - S$ only if there is a directed path from x_S to x_T in the network.
2. To observe x_T from x_S, we need to reconstruct $N_S = \sum_{n_i \subset S_i} n_i$ metabolite concentrations, where S denotes the set of all strongly connected components that are reachable from x_S, and n_i is the size of the i-th strongly connected component.
3. To identify the optimal sensor node for any target node, we can minimize N_s.

---

[1] We assume that the fluxes follow a mass action kinetics.

3.5 Exercises

Exercise 3.1 Prove for continuous-time systems, the following conditions are equivalent:

1. $x(0)$ is unobservable in time T.
2. $x(0)$ is unobservable in any time.

Exercise 3.2 Consider the condition

$$\text{rank} \left(\frac{\lambda I - A}{c^T} \right) = n, \quad \forall \lambda \in \mathbb{C}^n \tag{3.18}$$

and say if it is a sufficient or a necessary and sufficient condition for the observability of a continuous non-linear system.

Exercise 3.3 Consider a time-continuous non-linear input-free system Σ, whose dynamics is described by the following general equations:

$$\Sigma : \quad \dot{x}(t) = f(t, x), \quad f : [t_0, T] \times E \subset \mathbb{R}^n \longrightarrow \mathbb{R}^n, \tag{3.19}$$

whose experimentally observed output is

$$y(t) = h(t, x), \quad h : [t_0, T] \times E \subset \mathbb{R}^n \longrightarrow \mathbb{R}^n. \tag{3.20}$$

Suppose that the k-th order derivatives of f and h exist for every $x \in E$ and every $t \in [t_0, T]$, and that $y(t)$ is smooth, so that we can approximate $y(t)$ by a truncated Taylor series as follows:

$$y(t) = y(t_0) + \dot{y}(t_0)(t - t_0) + \frac{\ddot{y}(t_0)}{2!}(t - t_0)^2 + \cdots + \tag{3.21}$$
$$+ \frac{y^{(k-1)}(t_0)}{(k-1)!}(t - t_0)^{k-1} + \frac{y^{(k)}(t^*)}{k!}(t - t_0)^k,$$

where $t^* \in (t_0, T)$, and

$$y(t_0) = h(x(t_0), t_0) \equiv h_0(x(t_0), t_0)$$
$$\dot{y}(t_0) = \frac{\partial h_0}{\partial t}(x(t_0), t_0) + \left(\frac{\partial h_0}{\partial x(t_0)}(x(t_0), t_0) \right) f(x(t_0), t_0) \equiv h_1(x(t_0), t_0)$$

$$\vdots$$

$$y^{(k-1)}(t_0) = \frac{\partial h_{k-2}}{\partial t}(x(t_0), t_0) + \left(\frac{\partial h_{k-2}}{\partial x(t_0)}(x(t_0), t_0) \right) f(x(t_0), t_0) \equiv h_{k-1}(x(t_0), t_0).$$
$$\tag{3.22}$$

These equations can be written as a non-linear map

$$\mathbf{z} = \mathbf{Q}(x(t_0)),$$ (3.23)

where

$$\mathbf{z} = \begin{bmatrix} y(t_0) \\ \dot{y}(t_0) \\ \vdots \\ y^{(n-1)}(t_0) \end{bmatrix}$$ (3.24)

and $\mathbf{Q}$, called the *observability mapping* of Σ, is

$$\mathbf{Q}(x(t_0)) = \begin{bmatrix} h_0(x(t_0), t_0) \\ h_1(x(t_0), t_0) \\ \vdots \\ h_{k-1}(x(t_0), t_0) \end{bmatrix}.$$ (3.25)

Prove that the system Σ is completely observable if $\mathbf{Q}$ is a one-to-one mapping.

Exercise 3.4 Consider the Rössler model

$$\begin{cases} \dot{x}_1 = -x_2 - x_3 \\ \dot{x}_2 = x_1 + ax_2 \\ \dot{x}_3 = b + x_3(x_1 - c) \end{cases} .$$ (3.26)

Assume that $y = x_2$ can be measured. Is the model observable from $y = x_2$? And from $y = x_1$?

Exercise 3.5 In proposing the model in Exercise 3.4, Rössler was motivated by the search for chemical chaos, that is, chaotic behaviour[2] far-from-equilibrium chemical kinetics. With Willamowski, Rössler proposed the following chemical reaction scheme [39]:

$$R_1 : A_1 + X \underset{k_{-1}}{\overset{k_1}{\rightleftharpoons}} 2X$$

$$R_2 : X + Y \underset{k_{-2}}{\overset{k_2}{\rightleftharpoons}} 2Y$$

$$R_3 : A_5 + Y \underset{k_{-3}}{\overset{k_3}{\rightleftharpoons}} A_2$$

$$R_4 : X + Z \underset{k_{-4}}{\overset{k_4}{\rightleftharpoons}} A_3$$

---

[2] A system is said chaotic if it is highly sensitive to the initial conditions.

$$R_5 : A_4 + Z \underset{k_{-5}}{\overset{k_5}{\rightleftharpoons}} 2Z.$$

Reactions R_1 and R_5 are two auto-catalytic steps involving the species X and Z coupled to another catalytic step, that is, reaction R_2 involving species Y.

1. Write a mass action law model of this system.
2. Show that the Rössler model given in Exercise 3.4 is a minimal model for the mass action law model of this system.

References

1. Chou I-C, Voit EO. Recent developments in parameter estimation and structure identification of biochemical and genomic systems. Math Biosci. 2009;219(2):57–83.
2. Goel G, Chou I-C, Voit EO. System estimation from metabolic time-series data. Bioinformatics. 2008;24(21):2505–11.
3. Voit EO, Almeida J, Marino S, Lall R, Goel G, Neves AR, Santos H. Regulation of glycolysis in lactococcus lactis: an unfinisched systems biological case study. IEE Proc-Syst Biol. 2006;153(4):286–98.
4. Polisetty PK, Voit EO. Identification of metabolic system parameters using global optimization methods. Theor Biol Med Model. 2006;3(4):1–15.
5. Rodrigez-Fernandez M, Mendes P, Banga J. A hybrid approach for efficient and robust parameter estimation in biochemical pathways. BioSystems. 2006;83:248–65.
6. Moles GC, Mendes P, Banga JR. Parameter estimation in biochemical pathways: a comparison of global optimization methods. Genome Res. 2003;13:2467–74.
7. Tian T, Xu S, Burrage K. Simulated maximum likelihood method for estimating kinetic rates in gene expression. Bioinformatics. 2007;23(1):84–91.
8. Chou IC, Martens H, Voit EO. Parameter estimation in biochemical systems models with alternating regression. Theor Biol Med Model. 2006;3:25.
9. Sugimoto M, Kikuchi S, Tomita M. Reverse engineering of biochemical equations from time-course data by means of genetic programming. BioSystems. 2005;80:155–64.
10. Reinker S, Altman RM, Timmer J. Parameter estimation in stochastic biochemical reactions. IEEE Proc Syst Biol. 2006;153:168–78.
11. Hoops S, Sahle S, Gauges R, Lee C, Pahle J, Simus N. Copasi - a complex pathway simulator. Bioinformatics. 2006;22:3067–74.
12. Zwolak JW, Tyson JJ, Watson LT. Estimating rate constants in cell cycle models. In: Tentner A, editor. Proceedings of high performance constants in cell cycle models, San Diego. 2001. p. 53–7.
13. Vyshemirsky V, Girolami MA. Biobayes: Bayesian inference for systems biology; 2008.
14. Vyshemirsky V, Girolami MA. Bayesian ranking of biochemical system models. Bioinformatics. 2008;24(6):833–9.
15. Vyshemirsky V, Girolami MA. Biobayes: a software package for bayesian inference in systems biology. Bioinformatics. 2008;24(17):1933–4.
16. Boys RJ, Wilkinson DJ, Kirkwood TB. Statistics and Computing., Bayesian inference for a discretely observed stochastic kinetic modelNetherlands: Springer; 2008.
17. Golightly A, Wilkinson DJ. Bayesian inference for nonlinear multivariate diffusion models observed with error computational statistics and data analysis. Computational Statistics and Data Analysis. 2008;52(3):1674–93.

18. Wilkinson D. Stochastic Modelling for Systems Biology. Boca Raton: Chapman and Hall/CRC; 2006.
19. Wilkinson DJ. Bayesian methods in bioinformatics and computational systems biology. Briefings in Bioinformatics. 2007;1(8):109–16.
20. Lecca P, Palmisano A, Ihekwaba A, Priami C. Calibration of dynamic models of biological systems with kinfer. Eur Biophys J.
21. Geffen D, Findeise R, Schliemann M, Allgoever F, Guay M. Observability based paramter identifiability for biochemial reaction network. In: 2008 American control conference; 2008. p. 2130–34, June 11–13 2008.
22. Anguelova M. Nonlinear observability and identifiability: general theory and a case study of a kinetic model of s. cerevisiae; 2004.
23. Denis-Vidal L, Joly-Blanchard G, Noiret C. Some effective approaches to check the identifiability of uncontrolled nonlinear systems. Math Comput Simul. 2001;57(2):35–44.
24. Geffen D. Parameter identifiability of biochemical reaction networks in systems biology. PhD thesis, Department of Chemical Engineering of Queen's University, Kingston, Ontario, Canada; 2008.
25. Ljung L, Glad T. On global identifiability for arbitrary model parametrization. Automica. 1994;30(2):265–76.
26. Pohjanpalo H. System identifiability based on the power series expansion of the solution. Mathematical Biosciences. 1978;41:21–33.
27. Liu Y-Y, Slotine J-J, Barabasi A-L. Observability of complex systems. Proc Natl Acad Sci. 2013;110(7):2460–5.
28. Villaverde Alejandro F. Observability and structural identifiability of nonlinear biological systems. Complexity. 2019;2019:1–12.
29. Stigter JD, Joubert D, Molenaar J. Observability of complex systems: finding the gap. Sci Rep. 2017;7(1).
30. Lin Wu, Li Min, Wang Jian-Xin, Fang-Xiang Wu. Controllability and its applications to biological networks. J Comput Sci Technol. 2019;34(1):16–34.
31. Anguelova M. Observability and identifiability of nonlinear systems with applications in biology. PhD thesis, Chalmers University of Technology and G?eborg University; 2007.
32. Lecca Paola, Re Angela. Identifying necessary and sufficient conditions for the observability of models of biochemical processes. Biophys Chem. 2019;254:106257.
33. Aguirre LA, Portes LL, Letellier C. Structural, dynamical and symbolic observability: from dynamical systems to networks. PLOS ONE. 2018;13(10):e0206180.
34. Hong wei Lou and Rong Yang. Necessary and sufficient conditions for distinguishability of linear control systems. Acta Math Appl Sinica Engl Ser. 2014;30(2):473–82.
35. Krener J. $(ad_{f,g})$, $(ad_{f,g})$ and locally $(ad_{f,g})$ invariant and controllability distributions. SIAM J Control Optim. 1985;23(4):523–49.
36. Kawano Yu, Ohtsuka Toshiyuki. Observability analysis of nonlinear systems using pseudolinear transformation. IFAC Proc Vol. 2013;46(23):606–11.
37. Halás M, Kawano Y, Moog CH, Ohtsuka T. Realization of a nonlinear system in the feedforward form: a polynomial approach. IFAC Proc Vol. 2014;47(3):9480–5.
38. Stephanopoulos Gregory. Metabolic fluxes and metabolic engineering. Metab Eng. 1999;1(1):1–11.
39. Gaspard PP. Rössler systems.

Chapter 4
Regression and Variable Selection

Abstract Regression is used for explaining the relationship between a variable Y called *response*, and a set of one or more variables $X_1, X_2, \ldots, X_N$ called *explanatory variables*. The variable selection is the process that choose a reduced number of explanatory variables to describe a response variable in a regression models. Variable selection is used to (i) make the model easier to interpret, removing redundant non-informative variables; (ii) reduce the size of the problem to enable algorithms to run faster; and (iii) reduce the overfitting and, consequently, make the model predictive.

4.1 Multi-linear Regression

The multiple linear regression equation is as follows:

$$\hat{Y} = \beta_0 + \beta_1 X_1 + \beta_2 X_2 + \cdots + \beta_n X_n + \varepsilon, \tag{4.1}$$

where $\hat{Y}$ is the predicted or expected value of the dependent variable (hereafter called *response*), X_1 through X_n are n distinct independent or predictor variables, β_0 is the value of $\hat{Y}$ when all of the independent variables (X_1 through X_n) are equal to zero, and β_1 through $\beta_n X$ are the estimated regression coefficients. ε, the error term, is an independent random variables with a normal distribution of mean 0 and variance σ^2 [1].

Each regression coefficient represents the change in $\hat{Y}$ relative to a one unit change in the respective independent variable. In the multiple regression situation, β_1, for example, is the change in $\hat{Y}$ relative to a one unit change in X_1, holding all other independent variables constant (i.e. when the remaining independent variables are held at the same value or are fixed). Statistical tests can be performed to assess whether each regression coefficient is significantly different from zero.

Multiple linear regression analysis makes several assumptions.

1. There must be a linear relationship between the dependent variable and the independent variables.
2. The residuals are normally distributed.

© The Author(s), under exclusive license to Springer Nature Switzerland AG 2020   49
P. Lecca, *Identifiability and Regression Analysis of Biological Systems Models*,
SpringerBriefs in Statistics, https://doi.org/10.1007/978-3-030-41255-5_4

3. There is not multicollinearity, i.e. the independent variables are not highly correlated with each other.
4. Homoscedasticity of the errors' variances, i.e. the variance of error terms is similar across the values of the independent variables. A plot of standardized residuals versus predicted values can show whether points are equally distributed across all values of the independent variables.

The computational pipeline of the multi-linear regression analysis consisted of the following steps.

1. Fit a multi-linear model inclusive of all putative predictors to the experimental data.
2. Estimate the p-values of the t-test on the regression coefficients.
3. Evaluate collinearity: Calculate the variance inflation index (VIF); the threshold is traditionally set to 5.
4. Check for non-linearities by inspecting the partial residual plots.
5. Estimate the errors' autocorrelation.

The mathematical foundations and the algorithmic procedure implementing steps (1), (2) and (5) are very well known to the majority of the practitioners also without a mathematical education. The steps (3) and (4) instead use less known analysis tools, although they are of particular importance in multi-linear regression diagnostic. They are especially useful in our case study where we use a multi-linear model and its diagnostics as (i) explorative tools for the identification of the best predictors and (ii) indicators of relationships other than linear. In the following, we give a definition of the VIF index and of the partial residual plots. We refer the reader to the work of Lecca et al. [2] from where some parts of this chapter are also taken.

Partial Residual Plots

When performing a linear regression with a single independent variable, a scatter plot of the response variable against the independent variable provides a good indication of the nature of the relationship. Usually, in order to have an indication of the nature of the relationship, scatter plots of the response variable against the independent variables are provided. However, in this single scatter plot, the effect of the other independent variables in the model is not taken into account. Partial residual plots (often called component plus residuals plots) instead show the relationship between a given independent variable and the response variable given that other independent variables are also in the model [3–6]. On a partial residual plot, we plot

$$\text{Residuals} + \beta_i X_i \quad \text{versus} \quad X_i,$$

where residuals are the residuals from the full model. The partial residual plot allows to show the relationship between a given explanatory variable and the response variable by excluding the influence of the other predictors of the model. Consequently, this kind of plot is capable to identify the predictors linked to the response by a non-linear function and suggests the relevant refinements of the model. For instance, suppose that the k-*th* predictor results non-linearly correlated to the response; then, if f is the best candidate function describing this non-linearity, the model has to be amended as

$$\hat{Y} = \beta_0 + \beta_1 X_1 + \beta_2 X_2 + \cdots + f(X_k) + \cdots \beta_n X_n + \varepsilon.$$

Variance Inflation Factor

The variances of the estimated coefficients are inflated when multicollinearity exists. So, the variance inflation factor for the estimated coefficient β_k is just the factor by which the variance is inflated [7]. For simplicity, suppose that X_k is the only predictor, then

$$\hat{Y}_i = \beta_0 + \beta_k X_{ik} + \varepsilon_i, \tag{4.2}$$

where $i = 1, \ldots n$ ranges over the number of predictors. It can be shown that the variance of the estimated coefficient β_k is

$$\text{Var}_{\min}(\beta_k) = \frac{\sigma^2}{\sum_{i=1}^{n}(x_{ik} - \bar{x}_k)}. \tag{4.3}$$

Suppose to have a model with correlated predictors:

$$\hat{Y}_i = \beta_0 + \beta_1 X_{i1} + \cdots + \beta_k X_{ik} + \ldots \beta_{n-1} X_{i,(n-1)} + \varepsilon_i.$$

If some of the predictors are correlated with the predictor X_k, then the variance of β_k is inflated, and becomes

$$\text{Var}(\beta_k) = \frac{\sigma^2}{\sum_{i=1}^{n}(x_{ik} - \bar{x}_k)} \times \frac{1}{1 - R_k^2}, \tag{4.4}$$

where R_k^2 is the R^2 value obtained by regressing the k-*th* predictor on the remaining predictors. The greater the linear dependence among the predictor X_k and the other predictors, the larger the R_k^2 value, the larger the variance of β_k. The variance inflation factor quantifies how much larger the variance of the coefficient β_k becomes in case of multicollinearity [7] and thus

$$\text{VIF}(\beta_k) = \frac{\text{Var}(\beta_k)}{\text{Var}_{\min}(\beta_k)} = \frac{1}{1 - R_k^2}. \tag{4.5}$$

A VIF of 1 means that there is no correlation among the k-*th* predictor and the remaining predictor variables, and hence the variance of bk is not inflated at all. The general rule of thumb is that VIFs exceeding 4 suggest further investigations, while VIFs exceeding 10 indicate serious multicollinearity requiring model correction [7].

4.1.1   A Case Study: A Multi-linear Model of Sweat Secretion Volumes in Cystic Fibrosis Patients

The approach of multi-linear regression is frequently used as a first exploratory method to select explicative variables in the process of model construction in many application domains [8]. The author recently proposed in [2] a multi-linear model to identify the factors influencing the average ratio between the volume of secretory sweat droplets induced by CFTR gene-dependent stimuli and the volume of those formed by CFTR gene independent sweating. CODE 1 in Chap. 5 is the R script that implements this analysis.

In this study [2], the candidate variables are the number of glands, the volume sweat droplets from single glands, the concentration of chloride in sweat and the expiratory volume of the patient. The best predictors may be indeed diagnostic predictors of the disease. Indeed, these outcomes will support complex diagnosis. Quantification of CFTR function in CF patients and its improvement will be detectable during CFTR-targeted therapies by changes of selected outcomes.

Here, following are the names of the variables of the model and their meanings. The response is the *Mean CM_ratio* (Y) and its candidate predictors (Xs) are as follows:

- *Mch_glands*: number of glands induced to produce sweat droplets at 10 min after methacholine (Mch)-sweating stimulation via intradermal injection (M phase).
- *Cktl_glands* (nl/gl/min): number of glands induced to secrete sweat droplets at 30 min after cocktail (Cktl)-sweating stimulation via intradermal injection.
- *CM_ratio_glands*: ratio between the total number of glands induced in Cktl phase with the total number of droplets formed in MCh phase.
- *Mch_rate* (nl/min): average of volume of sweat droplets formed in MCh phase.
- *Cktl_rate*: average of volume of sweat glands formed in C phase.
- *FEV1* (forced expiratory volume in the 1st second): the volume of air that can be forced out in one second after taking a deep breath.
- *Sweat_Cl* (mmol/L): concentration of chloride that is excreted in sweat.
- *Gender* of the patient.

Mean CM_ratio has been computed as ratio between CFTR-dependent, evoked by intradermal microinjection of a β-adrenergic cocktail (Cktl), and CFTR-independent, induced by methacholine as cholinergic stimulus (MCh), sweat secretion rates by multiple individual glands. A number N of patients undergoing different pharmacological treatments are shown in Table 4.1.

We used the model to identify the set of explicative predictors in all the groups reported in the first columns of Table 4.1. The results of the analysis are summarized by Figs. 4.1, 4.2, 4.3, 4.4 and 4.5. We found that while in the control group (CTR) and in the group of heterozygotes (HTZ) the set of best predictors is the same and the multi-linear model fits quite well the observations, in the group of patients affected by CF, the set of the best predictors is different. Cktl_rate is the predictor present in both groups, but while in CTR and HTZ patients it correlates linearly with the response,

Table 4.1 Pharmacological treatments and number of tested patients [2]

Condition/treatment	Meaning	N
Kalydeko	ivacaftor (CFTR modulator)	2
Orkambi	ivacaftor+lumacaftor (CFTR modulator)	38
PTC124	Ataluren, PTC124 (CFTR modulator)	17
CRD	CFTR related disorder	9
UNK	Controversial diagnosis	8
DCP	Primary ciliary dyskinesia	5
BPCO	Bronchial pulmonary chronic Obstructive disease	4
CF	Cystic fibrosis	59
CTR	Control	32
HTZ	Heterozygous	33
PTC T0 + Orkambi T0	Before treatment with Ataluren or Orkambi	27

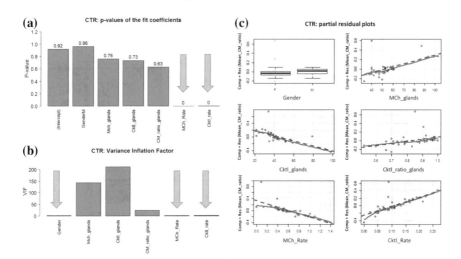

Fig. 4.1 Control group. **a** p-values of the fit coefficients; **b** VIF for each fit coefficient; **c** partial residual fits for each fit coefficient. The green arrows indicate significant predictors and low-VIF values fit coefficients

in the group of CF patients, it exhibits a non-linear behaviour (see Table 4.2 and Fig. 4.3c). Finally, we found that the multi-linear model does not properly describe the response. The two pharmacological treatments (Orakambi and PTC 124) induces non-linear behaviours of the response as function of most of the candidate predictors.

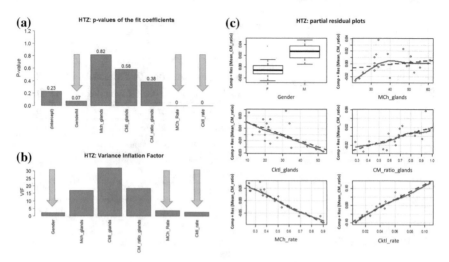

Fig. 4.2 Heterozygous group. **a** p-values of the fit coefficients; **b** VIF for each fit coefficient; **c** partial residual fits for each fit coefficient. The green arrows indicate significant predictors and low-VIF values fit coefficients

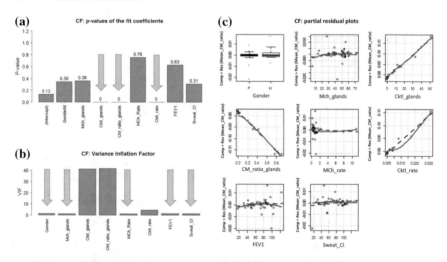

Fig. 4.3 Cystic fibrosis group. **a** p-values of the fit coefficients; **b** VIF for each fit coefficient; **c** partial residual fits for each fit coefficient. The green arrows indicate significant predictors and low-VIF values fit coefficients

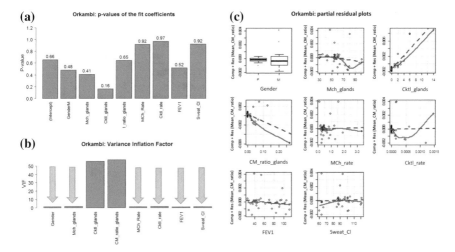

Fig. 4.4 Orkambi-treated group. **a** p-values of the fit coefficients; **b** VIF for each fit coefficient; **c** partial residual fits for each fit coefficient. The green arrows indicate significant predictors and low-VIF values fit coefficients

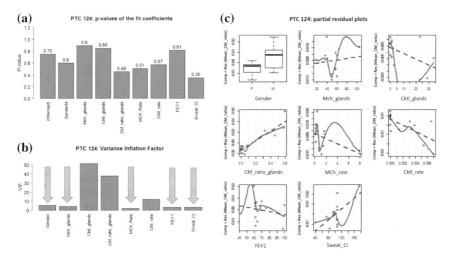

Fig. 4.5 PTC-treated group. **a** p-values of the fit coefficients; **b** VIF for each fit coefficient; **c** partial residual fits for each fit coefficient. The green arrows indicate significant predictors and low-VIF values fit coefficients

Table 4.2 The response Mean_CM_ratio in different groups is described by different sets of predictors

Group	Independent predictors VIF < 5
CTR	Gender, MCh_rate, Cktl_rate
HTZ	Gender, MCh_rate, Cktl_rate
CF	Gender, MCh_rate, Cktl_rate
Orlambi	Gender, MCh_glands, MCh_rate, Cktl_rate, FEV1, Sweat_Cl
PTC 124	Gender, MCh_glands, MCh_rate, FEV1, Sweat_Cl

Group	Best predictors p-value < 10%s	Nonlinearities partial residual plots
CTR	MCh_rate, Cktl_rate	–
HTZ	MCh_rate, Cktl_rate	MCh_glands
CF	Cktl_glands, CM_ratio_glands, Cktl_rate	Cktl_rate
Orlambi	Cktl_glands	MCh_glands, Cktl_rate, CM_ratio_glands
PTC 124	–	All the predictors

4.2 Robust Regression

Robust regression is recommended specifically when the data contain outliers, whose presence violates the assumption of normally distributed residuals and distorts estimate of the least squares coefficients by having more influence than the other data. In a dataset with N data devoid of outliers, the weight of each observation is $1/N$. However, outlying observations may receive a weight of 10, 20 or even 50%. The increase of the weights corresponding to the outliers is the primary cause of the distortion in the regression coefficient estimates. In turn, because of this distortion, these outliers are difficult to identify since their residuals are much smaller than they should be. Robust regression down weights the influence of the outliers by making their residuals larger, so that their impact on the coefficient estimates is minimized.

There exist several families of robust estimators ([9–11]). Here, we mention the *M-estimators*, which is implemented in the majority of the software for robust regression. The M-estimator minimizes the sum of a function ρ of the residuals:

$$\min_{\beta} \sum_{i=1}^{N} \rho(y_i - x_i^{\top}\beta) \equiv \min_{\beta} \sum_{i=1}^{N} \rho(e_i). \tag{4.6}$$

Note that if the residuals follow the normal distribution, and if $\rho(u) = \frac{1}{2}u^2$, then minimizing ρ results in minimizing the least squares method. For this reason, "M" in M-estimators stands for maximum likelihood. In order to make M-estimators scale invariant, a scale factor s is introduced in this way

$$\min_{\beta} \sum_{j=1}^{n} \rho\left(\frac{y_j - x_j^\top \beta}{s}\right) = \min_{\beta} \sum_{j=1}^{n} \rho\left(\frac{e_j}{s}\right). \tag{4.7}$$

s can be found by solving equation (4.7). For instance, in NCSS statistical software [12], the estimate of s is

$$s = \frac{\text{median}|e_i - \text{median}(e_j)|}{0.6745}, \tag{4.8}$$

and when N is large and the error is normally distributed, it is an approximate estimator of the standard deviation of the residuals. To minimize the function $\sum_{j=1}^{n} \rho\left(\frac{y_j - x_j^\top \beta}{s}\right)$, the first partial derivatives of $\rho(\cdot)$ with respect to the β_i $(i = 1, 2, \ldots, p)$ are set to zero. In this way, a set of $p + 1$ non-linear equations is obtained

$$\sum_{j=1}^{n} x_{ij}\varphi\left(\frac{y_j - \beta x_j}{s}\right) = 0, \tag{4.9}$$

where $\varphi(u) = \rho'(u)$ is called *influence function* [12]. To solve these equations, *iterative reweighted techniques* [13] are used. At the first iteration, the ordinary least squares regression coefficients are used to begin the iterative procedure. At each step, new estimates of the regression coefficients are found using the matrix equation

$$\beta_{k+1} = (X^\top W_k X)^{-1}(X^\top W_k Y), \tag{4.10}$$

where W_k is a $N \times N$ diagonal matrix of weights $w_{1k}, w_{2k}, \ldots, w_{Nk}$ defined as follows:

$$\begin{cases} w_{jk} = \dfrac{\varphi\left[\frac{(y_j - x^\top \beta_{jk})}{s_k}\right]}{\frac{y_j - x^\top \beta_{jk}}{s_k}} & \text{if } y_j \neq x^\top \beta_{jk} \\ \qquad\qquad 1 & \text{if } y_j = x^\top \beta_{jk}, \end{cases}$$

The Huber's method and the Tukey's biweight defines the functions $\rho(\cdot)$, $\varphi(\cdot)$ and $w(\cdot)$ as follows.

Huber's Method

$$\rho(u) = \begin{cases} u^2 & \text{if } |u| < \varepsilon \\ 2|u|\varepsilon - \varepsilon^2 & \text{if } |u| \geq \varepsilon \end{cases}$$

$$\varphi(u) = \begin{cases} u \text{ if } |u| < \varepsilon \\ \varepsilon \text{ sign}(u) \text{ if } |u| \geq \varepsilon \end{cases}$$

$$w(u) = \begin{cases} 1 \text{ if } |u| < \varepsilon \\ \frac{c}{|u|} \text{ if } |u| \geq \varepsilon \end{cases}$$

with $\varepsilon = 1.345$ [12].

Tukey's Biweight

$$\rho(u) = \begin{cases} \frac{c^3}{3}\left\{ 1 - \left[1 - \left(\frac{u}{\varepsilon}\right)^2 \right]^3 \right\} \text{ if } |u| < \varepsilon \\ 2\varepsilon \text{ if } |u| \geq \varepsilon \end{cases}$$

$$\rho(u) = \begin{cases} u\left[1 - \left(\frac{u}{\varepsilon}\right)^2 \right]^2 \text{ if } |u| < \varepsilon \\ 0 \text{ if } |u| \geq \varepsilon \end{cases}$$

$$w(u) = \begin{cases} u\left[1 - \left(\frac{u}{\varepsilon}\right)^2 \right]^2 \text{ if } |u| < \varepsilon \\ 0 \text{ if } |u| \geq \varepsilon \end{cases}$$

with $\varepsilon = 4.685$ [12].

4.3   Variable Selection in Non-linear Regression

The variable selection problem also occurs in non-linear models. Since, in some situations, the functional form of the model may be unknown,[1] the problem of identifying a subset of explicative variables, if it exists, is then of great importance. The majority of methods for variable selection methods rely on linear models, but in the last two decades fewer, but effective sophisticated solutions have been proposed also for non-linear models [14–19].

A simple method consists in using mutual information which can capture both linear and non-linear dependencies. The mutual information have to be calculated between the candidate predictors and the response variable and based on this criterion select relevant features.

There is a family of simple and famous models that captures both linear and non-linear interactions. The most known representative of this family is the *decision tree*, which is also very popular in biological and medical applicative domains [20–25]. The algorithm of the decision tree models implements a recursive partitioning of the data, i.e. it repeatedly partitions the data into multiple subspaces, so that the

---

[1]CODE 3 shows a R script implementing an unsupervised model selection.

outcomes in each final subspace are as homogeneous as possible. The result consists of a set of rules used for predicting the outcome variable, which can be either a continuous variable, for *regression trees*, or a categorical variable, for *classification trees*. The decision rules generated by the CART (Classification and Regression Trees) predictive model are generally visualized as a binary tree.

Here, we will focus on regression trees. Basic regression trees partition a data set into smaller groups and then fit a constant line for each subgroup. Unfortunately, a single tree model tends to be highly unstable and a poor predictor. By bootstrap aggregating (bagging) regression trees, this technique, that for a single tree is highly unstable and a poor predictor, can become powerful and effective. Regression tree, originally developed [26], provides the fundamental basis of more complex tree-based models such as *random forests* and *gradient boosting machines*.

Basic regression trees partition a dataset into smaller subgroups and then fit a simple constant for each observation in the subgroup. The partitioning is achieved by successive binary partitions (recursive partitioning) based on the different predictors. The constant to predict is based on the average response values for all observations that fall in that subgroup. Formally, let Y be the response variable and $X_1, X_2, \ldots, X_n$ be the n candidate predictors. The recursive partitioning results in m regions $R_1, R_2, \ldots, R_m$, where the model predicts with a constant line c_i for region R_i:

$$Y(X_1, X_2, \ldots, X_n) = \sum_{i=1}^{n} c_i I(X_1, X_2, \ldots, X_n). \qquad (4.11)$$

The partitioning of variables is implemented by a top-down, greedy algorithm, i.e. a partition performed earlier in the tree will not change based on later partitions. The model begins with the entire dataset, S, and searches every distinct value of every candidate predictor to find the explicative predictor and split value that partitions the data into two regions (R_a and R_b) such that the overall sums of squares error are minimized:

$$\min \left\{ SSE = \sum_{j\,in\,R_a} (y_j - c_a)^2 + \sum_{j\,in\,R_b} (y_j - c_b)^2 \right\}. \qquad (4.12)$$

Having found the best split, the data are partitioned into the two resulting regions and the splitting process is repeated on each of the two regions. This process is continued until some stopping criterion is reached. What results is, typically, a very deep, complex tree that may produce good predictions on the training set but is likely to overfit the data, leading to poor performance on new data. There is often a balance to be achieved in the depth and complexity of the tree to optimize predictive performance on some new data. To find this balance, a very large tree is generated as defined and then pruned it back to find an optimal subtree. The optimal subtree is pruned by using a cost complexity parameter (α) that penalizes our objective function in Eq. (4.12) for the number of terminal nodes of the tree (T) as in the following equation:

$$\min \left\{ \mathrm{SSE} + \alpha |T| \right\}. \tag{4.13}$$

For a set value of α, the smallest pruned tree that has the lowest penalized error is returned. The fit of a decision tree returns a predictors' importance vector that will tell report the relevance of each predictor.

Here, we gave a short summary of regression tree methodology to select explicative predictors. Good regression tree tutorials, where the reader can deepen the regression tree methodology, can be found, for example, in [27, 28].

4.4 Exercises

Exercise 4.1 We want to estimating how wage income is determined by education and working experience. A well-known example is Card (1993) [29] who investigates the economic return to schooling and uses college proximity as an instrumental variable.

Consider in R the dataset `CollegeDistance` which is similar to the data used by Card (1993). It stems from a survey of high school graduates with variables coded for wages, education, average tuition and a number of socio-economic measures. The dataset also includes the distance from a college, while the survey participants were in high school. `CollegeDistance` comes with the `AER` package.

1. Determine if this is a linear or a non-linear regression problem.
2. Implement a suitable variable selection method to build a model of wage income.

Exercise 4.2 Consider in R the dataset `Concrete` which contains data from an experiment testing the properties of concrete made with different proportions of ingredients. The variables of interest are compressive strength of the concrete after 28 days, the percentage of limestone, and water–cement ratio.

1. Verify if the assumption of the linear regression is satisfied.
2. If the assumptions for linear regression are satisfied, implement a multi-linear regression model to see how compressive strength depends on the "recipe" of ingredients; otherwise, implement an alternative suitable model.

Exercise 4.3 Consider in R the `crime` dataset, including the following variables: state id (`sid`), state name (`state`), violent crimes per 100,000 people (`crime`), murders per 1,000,000 (`murder`), the percent of the population living in metropolitan areas (`pctmetro`), the percent of the population that is white (`pctwhite`), percent of population with a high school education or above (`pcths`), percent of population living under poverty line (`poverty`) and percent of population that are single parents (`single`). The dataset has 51 observations. Use poverty and single to predict crime.

1. Verify if the assumption of the linear regression is satisfied.
2. Say if traditional regression can be applied, or if robust regression is more suitable for this case study. Motivate the answer.
3. If robust regression is considered more appropriate for this case study, implement it.

Exercise 4.4 Consider in R the `iris` data frame.

1. Which variables are explicative of the variable `Species`?
2. Build a tree to classify `Species` from the other variables.
3. Plot the tree.

Hint: Use the function `rpart` function of the library `rpart`.

Exercise 4.5 Explain why

1. incorrect signs of the regression coefficients may indicate multicollinearity;
2. instability of the regression coefficients may indicate multicollinearity.

Hint: to test for instability of the coefficients, run the regression on different combinations of the variables and see how much the estimates change.

Exercise 4.6 CODE 3 in Chap. 5 implements the Jacquelin's [30] method for the regression of a two-term sum of exponentials

$$y(x) = ae^{-px} + be^{-qx}, \quad p, q > 0$$

1. Complete the code by adding to it statistical tests to assess the significance of the parameter estimates a, b, p and q.
2. Suppose $X_1, X_2, \ldots, X_n$ is a set of candidate predictors of Y and that the model is

$$y(x_1, x_2, \ldots, x_n) = a_0 + \sum_{i=1}^{n} a_i e^{p_i x_i}.$$

Design a method to select the most explicative predictors, and implement it to proved an extension of the CODE 3.

References

1. Garson GD. Multiple regression statistical. Publishing Associates; 2014.
2. Lecca P, Bertini M, Vicentini R, Pintani E, D'Orazio C, Esposito V, Kleinfelder K, Sorio C, Melotti P. Multilinear regression analysis of sweat secretion volumes in cystic fibrosis patients. In Proceedings of the 23rd conference of open innovations association FRUCT, FRUCT'23. FRUCT Oy: Helsinki, Finland; 2018, p. 68:497–68:504.
3. Draper NR, Smith H. Applied regression analysis. New Jersey: Wiley; 1998.
4. Neter J, Wasserman W, Kunter MH. Applied Linear Statistical Models. Irwin; 1990.

5. McKean JW, Sheather SJ. L_1-Statistical procedures and related topics, Chapter exploring data sets using partial residual plots based on robust fits, vol. 31; 1997.
6. Ryan T. Modern regression methods. New Jersey: Wiley; 1997.
7. Web page of the online courses of pennstate eberly college of science detecting multicollinearity using variance inflation factors.
8. Nimon KF. Statistical assumptions of substantive analyses across the general linear model: a mini-review. Front Psychol. 2012;3.
9. Wilcox RR. Introduction to robust estimation and hypothesis testing (Statistical modeling and decision science). Cambridge: Academic; 2016.
10. Pitselis G. A review on robust estimators applied to regression credibility. J Comput Appl Math. 2013;239:231–49.
11. Zaman T, Bulut H. Modified ratio estimators using robust regression methods. Commun Stat-Theory Methods. 2018;48(8):2039–48.
12. NCSS Statistical Software. https://www.ncss.com. Accessed 01 Sept 2019.
13. Xiu X, Kong L, Li Y, Qi H. Iterative reweighted methods for $\ell\_1 - \ell\_p\ell 1 - \ell p$ minimization. Comput Optim Appld. 2018;70(1):201–19.
14. Rech G, Teräsvirta T, Tschernig R. A simple variable selection technique for nonlinear models. Wirtschaftswissenschaftliche Fakultät: Humboldt-Universität zu Berlin; 1999.
15. Peduzzi PN, Hardy RJ, Holford TR. A stepwise variable selection procedure for nonlinear regression models. Biometrics. 1980;36(3):511.
16. Gregorová M, Kalousis A, Marchand-Maillet S. Structured nonlinear variable selection; 2018.
17. Deng H, Runger G. Feature selection via regularized trees. In: The 2012 international joint conference on neural networks (IJCNN). IEEE; 2012.
18. Benoît F, van Heeswijk M, Miche Y, Verleysen M, Lendasse A. Feature selection for nonlinear models with extreme learning machines. Neurocomputing. 2013;102:111–24.
19. Wu S, Xue H, Wu Y, Wu H. Variable selection for sparse high-dimensional nonlinear regression models by combining nonnegative garrote and sure independence screening. Stat Sin. 2014.
20. Bode-Greuel KM, Greuel JM. Determining the value of drug development candidates and technology platforms. J Commer Biotechnol. 2005;11(2):155–170.
21. Geurts P, Irrthum A, Wehenkel L. Supervised learning with decision tree-based methods in computational and systems biology. Mol BioSyst. 2009;5(12):1593.
22. Dimitris D. Healthcare biotechnology: a practical guide. Boca Raton: CRC Press; 2010.
23. Djuris J, Ibric S, Djuric Z. Neural computing in pharmaceutical products and process development. In: Computer-aided applications in pharmaceutical technology. Amsterdam: Elsevier; 2013, p. 91–175.
24. Grudzinskas C, Gombar CT. Portfolio and project planning and management in the drug discovery, evaluation, development, and regulatory review process. In: Principles of clinical pharmacology. Amsterdam: Elsevier; 2012, p. 487–506.
25. Luna JM, Gennatas ED, Ungar LH, Eaton E, Diffenderfer ES, Jensen ST, Simone CB, Friedman JH, Solberg TD, Valdes G. Building more accurate decision trees with the additive tree. Proc Natl Acad Sci. 2019;116(40):19887–93.
26. Breiman L, Friedman J, Stone CJ, Olshen RA. Classification and regression trees. The Wadsworth and Brooks-Cole statistics-probability series. Taylor & Francis, Milton Park; 1984.
27. Regression trees. http://uc-r.github.io/regression_trees. Accessed 02 Nov 2019.
28. Tree-based models. https://www.statmethods.net/advstats/cart.html. Accessed 02 Nov 2019.
29. Card D. Using geographic variation in college proximity to estimate the return to schooling. NBER working papers 4483, National Bureau of Economic Research, Inc.
30. Régressions et équations intégrales. https://it.scribd.com/doc/14674814/Regressions-et-equations-integrales. Accessed 02 Nov 2019.

Chapter 5
R Scripts

Abstract In this chapter, we report the R scripts implemented to perform identifiability analysis and regression, mentioned in the previous chapters. The input files are only available upon request to the author.

5.1   CODE 1: Multi-linear Regression

Code 1 implements the analyses reported in Chap. 4, Sect. 4.1.1.

```
require(xlsx)
require(corrplot)
library(MASS)
library(car)
library(leaps)
require(dplyr)
library(ggfortify)


setwd(dirname(rstudioapi::getActiveDocumentContext()$path))

######################### FUNCTIONS ############################

# Function to detect which colums contains only NA values
# this colums have to be excluded from the multilinear fit.

ColNums_NotAllMissing <- function(df){ # helper function
as.vector(which(colSums(is.na(df)) != nrow(df)))
}

ColNums_NOTSeverallMissing <- function(df){ # helper function
as.vector(which(colSums(is.na(df)) <= round(nrow(df)/2),1))
}

# count the NaN in  vector
nacount <- function(x){
                        c <- 0
                        for(i in 1:length(x))
                          {
                            if(is.na(x[i]))
                              {
                                c <- c +1
```

```
                                              }
                                        }
                                   c
                             }

#####################################################################

######################### INPUT DATA #############################

input.data <- read.xlsx("./Cystic_Fibrosis.xlsx",
                          sheetName = "Foglio1", header=T)[,1:15]

# Subsetting according to the number of mutations (0, 1, and 2)
zero_mutations <-
subset(input.data, input.data[,
                    "Mutations"]==0)[,1:(dim(input.data)[2]-1)]

one_mutations <-
subset(input.data, input.data[,
                    "Mutations"]==1)[,1:(dim(input.data)[2]-1)]

two_mutations <-
subset(input.data, input.data[,
                    "Mutations"]==2)[,1:(dim(input.data)[2]-1)]

# remove colums with all the elements equal
zero_mutations <-
zero_mutations[vapply(zero_mutations,
                function(x) length(unique(x)) > 1, logical(1L))]

one_mutations <-
      one_mutations[vapply(one_mutations,
        function(x) length(unique(x)) > 1, logical(1L))]

two_mutations <-
      two_mutations[vapply(two_mutations,
        function(x) length(unique(x)) > 1, logical(1L))]


# size of each group
dim(zero_mutations)
dim(one_mutations)
dim(two_mutations)

# detect colums containing several NA values and remove them
zero_mutations <-
    zero_mutations %>% select(ColNums_NOTSeverallMissing(.))
one_mutations <-
    one_mutations %>% select(ColNums_NOTSeverallMissing(.))
two_mutations <-
    two_mutations %>% select(ColNums_NOTSeverallMissing(.))

colnames(zero_mutations)
colnames(one_mutations)
colnames(two_mutations)

# eliminate the ID columns (not necessary for the analysis)
zero_mutations <-  as.data.frame(zero_mutations[,-2])
one_mutations <- as.data.frame(one_mutations[,-2])
```

```
two_mutations <- as.data.frame(two_mutations[,-2])

# eliminate CFTR_Genotype if present
zero_mutations <-
zero_mutations[ , !names(zero_mutations) %in% c("CFTR_Genotype")]
one_mutations <-
one_mutations[ , !names(one_mutations) %in% c("CFTR_Genotype")]
two_mutations <-
two_mutations[ , !names(two_mutations) %in% c("CFTR_Genotype")]

# eliminate  Mean_CM_Ratio_Percen
zero_mutations <- zero_mutations[ , !names(zero_mutations)
                    %in% c("Mean_CM_Ratio_Percent")]
one_mutations <- one_mutations[ , !names(one_mutations)
                    %in% c("Mean_CM_Ratio_Percent")]
two_mutations <- two_mutations[ , !names(two_mutations)
                    %in% c("Mean_CM_Ratio_Percent")]


linear_model <- function(input.db, outdir1, outdir2)
{
lev <- levels(input.db[,1])
group <- list()
length(group) <- length(lev)

for (i in 1:length(lev))
{
# select groups according to different treatment
group[[i]] <- subset(input.db, input.db[,1] == lev[i])

group[[i]]<- group[[i]][ , !names(group[[i]]) %in% c("Group")]
print(dim(group[[i]])[1])

if( dim(group[[i]])[1] >=10)
   {
     # multilinear model
     mod.lm <- lm(Mean_CM_Ratio ~ ., data=group[[i]])

     # Evaluate Collinearity
     inflation <- vif(mod.lm) # variance inflation factors

     if(nacount(inflation) == 0)
       {
         png(paste(outdir2, "/",lev[i],"_VIF.png", sep=""),
             width=880, height=550)
         par(mar=c(15.1,5.1,9.1,2.1))
         barplot(inflation, ylab="VIF",
         main=paste(lev[i],": Variance Inflation Factor", sep=""),
          cex.names=1.5, cex.axis=1.7, cex.main=2, cex.lab=1.8,
          las=2, col="salmon")
         dev.off()
     }

# Evaluate Nonlinearity
# component + residual plot
png(paste(outdir2, "/",lev[i],"_Nonlinearity.png", sep=""))
crPlots(mod.lm, ylab="Compp. + res. (Mean_CM_Ratio)",
        main=paste(lev[i], ": partial residual plots", sep=""))
```

```
dev.off()

# Ceres plots
ceresPlots(mod.lm)

# Test for Autocorrelated Errors
sink(paste(outdir2, "/", lev[i], "_AutocorrErros.txt", sep=""))
print(durbinWatsonTest(mod.lm))
sink()

# Evaluate homoscedasticity
# non-constant error variance test
ncvTest(mod.lm)
plot studentized residuals vs. fitted values
png(paste(outdir2, "/",lev[i],"_Homoscedasticy.png", sep=""))
spreadLevelPlot(mod.lm)
dev.off()

# Global test of model assumptions
sink(paste(outdir2, "/", lev[i], "_Assumptions.txt", sep=""))
gvmodel <- gvlma(mod.lm)
summary(gvmodel)
sink()

# diagnostic plots (not always possible to draw the qqplot)
png(paste(outdir2, "/",lev[i],"_DIAGNOSTIC.png", sep=""))
par(mfrow = c(2, 2))
plot(mod.lm, cex.lab=1.2, cex.axis=1.2)
title(paste(lev[i], ": regression diagnostics", sep=""),
      line=-1.5, cex.main=1.5, outer=TRUE)
dev.off()


# stepwise regression for model selection

step <- stepAIC(mod.lm, direction="both")
step$anova # display results

leaps <- regsubsets(Mean_CM_Ratio ~ Gender + MCh_glands + Cktl_glands +
 CM_ratio_glands + MCh_Rate + Cktl_Rate + Mean_CM_Ratio_Star +
 CFTR_Genotype + FEV + Lung_infection + Sweat_Cl,
 data=input.df,nbest=3, really.big=T)

# view results
summary(leaps)

# plot a table of models showing variables in each model.
# models are ordered by the selection statistic.
plot(leaps,scale="r2")
plot statistic by subset size
subsets(leaps, statistic="rsq")

png(paste(outdir1, "/", lev[i],".png", sep=""), width=880, height=650)
par(mar=c(15.1,5.1,9.1,2.1))
xx <- barplot(summary(mod.lm)$coefficients[ ,4], las=2,
              main=paste(lev[i], ": p-values of the fit coefficients", sep=""),
              ylim=c(0, 1.2), col="salmon", ylab="P-value",
              cex.names=1.5, cex.lab=1.8, cex.axis=1.7, cex.main=2)
text(x = xx, y = summary(mod.lm)$coefficients[ ,4],
```

```
label = round(summary(mod.lm)$coefficients[ ,4], 2), pos = 3, cex = 1.6,
col = "black")
dev.off()

sink(paste(outdir1, "/", lev[i], "_lm.txt", sep=""))
print(summary(mod.lm))
sink()
}
}
}

linear_model(zero_mutations, "Model_PValues_Zero", "Diagnostics_Zero")
linear_model(one_mutations, "Model_PValues_Uno", "Diagnostics_Uno")
linear_model(two_mutations, "Model_PValues_Due", "Diagnostics_Due")
```

5.2   CODE 2: Unsupervised Regression

The following program (`Fit_data.R`) finds without the need of the user supervision the best function fitting experimental observations. An example of input file is shown in Table 5.1.

To run the program without executing new points imputation (implemented by a spline calculation), launch the command

```
Rscript --vanilla Fit_data.R -f input_data.xlsx
```

To run the program executing also new points imputation, launch the command

```
Rscript --vanilla Fit_data.R -f input_data.xlsx \{i Y
```

```
###########################################################
# Run: Rscript --vanilla Fit_data.R -f input_file.xlsx    #
#                                                         #
# Input file contains n columns and m rows. Each column   #
```

Table 5.1 Example of input file to `Fit_data.R`. It contains five experimental replicates of fluorescence measurements in a chemical mixture in response to different concentrations of iodine ($[I]$), in a typical fluorescence assay experiment with enzymatic reactions

$[I]$	Exp 1	Exp 2	Exp 3	Exp 4	Exp 5
0.7	0.76	0.78	0.79	0.81	0.79
1.5	0.67	0.72	0.71	0.62	0.66
3	0.52	0.55	0.55	0.49	0.5
5.9	0.36	0.37	0.37	0.3	0.34

```
# is an experimenta replicate.                                    #
# This is used to find the best fitting function to the           #
# curve of fluorescence against concentration of a stimulus. #
################################################################

wd <- getwd()
setwd(wd)

# Function to Install and Load R Packages
Install_And_Load <- function(Required_Packages)
{
    Remaining_Packages <- Required_Packages[!(Required_Packages
                %in% installed.packages()[,"Package"])];

    if(length(Remaining_Packages))
    {
        suppressMessages(suppressWarnings(
          install.packages(Remaining_Packages,
          repos="https://cloud.r-project.org")));
    }
    for(package_name in Required_Packages)
    {
        suppressMessages(suppressWarnings(
          library(package_name,
            character.only=TRUE,quietly=TRUE)));
    }
}

# Specify the list of required packages to be installed and load
Required_Packages=c("Hmisc", "XLConnect", "broom", "xlsx",
                    "minpack.lm", "optparse");

# Call the Function
Install_And_Load(Required_Packages);




################################################################################
#                              FUNCTIONS                                       #
################################################################################

# sum matrix columns
sumcol <- function(mat)
{
    somma <- array(0, dim(mat)[1])
    for (j in 1: dim(mat)[2])
    {
        somma <- somma + mat[,j]
    }
    somma
}


################################################################################

option_list = list(
make_option(c("-f", "--file"), type="character", default=NULL,
help="Input data Excel file", metavar="character"),
```

```
make_option(c("-i", "--spline"), type="character", default="N",
help="Performs cubic spline interpolation
        of given data [default= %default]", metavar="character")
);

opt_parser = OptionParser(option_list=option_list);
opt = parse_args(opt_parser);

if (is.null(opt$file)){
    print_help(opt_parser)
    stop("At least one argument must be supplied (input file).n",
        call.=FALSE)
}

interp <- as.character(opt$spline)

cat("\n")
cat("=====================================\n")
cat("\n Fit_data.R is an R script to estimate
cat("\n=====================================\n")

cat("\n Usage: \n")
cat("\n  Options:
-f CHARACTER, --file=CHARACTER
Input data (Excel file format): first column (x-axis),
other columns (replicates of the response variable (y))

-interp CHARACTER, --spline=CHARACTER
Performs cubic spline interpolation of given data [default= N]

-h, --help
Show this help message and exit
\n\n")
cat("\n=====================================\n")


# fit of fluorescence
process_fluorescence <- function(filein){

    input.data <- read.xlsx(filein, 1)
    mean.fluo <- sumcol(input.data[2:dim(input.data)[2]])
                        /(dim(input.data)[2] -1)

    if (interp=="Y")
    {
        x <- input.data[,1]
        y <- mean.fluo

        cat("\nCubic spline interpolation ... \n")
        interp.curve <- spline(x, y)
        input.data <- data.frame(NaI=interp.curve$x,
                                Fluorescence=interp.curve$y)
        # n.points <- length(input.data$NaI)
    }

    if (interp=="N")
    {
        input.data <- data.frame(NaI=input.data[,1],
                                Fluorescence=mean.fluo)
```

```
    n.points <- length(input.data$NaI)
}

cat ("\nPlease wait: the program is choosing
          the best fitting function ...\n")

# non-linear models 1: hyperbole
cat ("\nBasic fit 1: the reciprocal of the predictor:
      y = [1/(a + x)] + c\n")
non_linear.model_1  <- nlsLM(input.data$Fluorescence ~
                            I(1/(a + b*input.data$NaI) + c),
data=input.data,
start = list(a = 0.00001, b = 0.0000001, c= 1))

tidy_model_1 <- tidy(non_linear.model_1)
pvalues <- as.vector(tidy_model_1[,5])
# n.signif <- length(which(pvalues <= 0.05))
cat("Coefficients:\n")
print(tidy_model_1)
cat("\n")

# mean sum of residuals
residuals_model_1 <- residuals(non_linear.model_1)
MSR_1 <- sum(residuals_model_1^2)/length(residuals_model_1)

cat("Average residual sum of squares: ")
print(MSR_1)
cat("\n")

A <- coef(non_linear.model_1)[1]
B <- coef(non_linear.model_1)[2]
C <- coef(non_linear.model_1)[3]

y.model <- 1/(A + B*input.data$NaI) + C

cat("Total residual sum of squares: ")
TSS <- (sum((y.model - mean(input.data$Fluorescence))^2))
print(TSS)
cat("\n")

SSR <- sum(residuals_model_1^2)
cat("R-squared: ")
R.square <- 1 - (SSR/TSS)
cat(R.square)

cat("\n\n")


# non linear models 2: exponential
cat ("\nBasic fit 2: the exponential of the predictor:
      y = exp(a*x) + b\n")
non_linear.model_2  <- nlsLM(input.data$Fluorescence ~
                            exp(a*input.data$NaI) + b,
                            data=input.data,
                            start = list(a = 1, b = 1))
                            tidy_model_2 <- tidy(non_linear.model_2)

# pvalues <- as.vector(tidy_model_2[,5])
```

```
cat("Coefficients:\n")
print(tidy_model_2)
cat("\n")

# mean sum of residuals
residuals_model_2 <- residuals(non_linear.model_2)
MSR_2 <- sum(residuals_model_2^2)/length(residuals_model_2)

cat("Average residual sum of squares: ")
print(MSR_2)
cat("\n")

A <- coef(non_linear.model_2)[1]
B <- coef(non_linear.model_2)[2]

y.model <- exp(A*input.data$NaI) + B

cat("Total residual sum of squares: ")
TSS <- (sum((y.model - mean(input.data$Fluorescence))^2))
print(TSS)
cat("\n")

SSR <- sum(residuals_model_2^2)
cat("R-squared: ")
R.square <- 1 - (SSR/TSS)
cat(R.square)

cat("\n\n")


# non linear models 3: polynomial
cat ("\nBasic fit 3: polynomial:
      y = a0 + Sum_i (a_i*x_i^i), i=1,2,...\n")
k <- 1

if (interp=="Y") {N <- 10}
if (interp=="N") {N <- n.points -1}

for (i in 1:N)
{
    model <- lm(input.data$Fluorescence ~ poly(input.data$NaI,i))
    p.m <- glance(model)$p.value
    pos <- vector()
    if(!is.na(p.m) && (p.m < 0.05))
    {
        pos[k] <- i
        k <- k+1
    }
}
degree <- pos[1]

if (!is.na(degree) && (degree >=1))
{
    if (degree > 1)
    {cat("\nWarning: the degree of fitting
    polynomial is greater than 1: there may be multiple roots!\n")}
    non_linear.model_3 <- lm(input.data$Fluorescence ~
                          poly(input.data$NaI,degree))
```

```r
    model.pv <- glance(model)$p.value # p-value on F statistics

    model.coeffs <- tidy(non_linear.model_3)

    cat("Coefficients:\n")
    print(tidy(non_linear.model_3))
    cat("\n")

    # mean sum of residuals
    residuals_model_3 <- residuals(non_linear.model_3)
    MSR_4 <- sum(residuals_model_3^2)/length(residuals_model_3)

    cat("Mean sum of squared residuals: ")
    print(MSR_3)
    cat("\n")
}

else {
    MSR_3 <- Inf
    cat("\nPolynomial fit are not adequate.\n\n")
}

MSRs <- c(MSR_1, MSR_2, MSR_3)

selected_model <- which.min(MSRs)

for (j  in 1:97)
{cat("-")}
cat("\n")

cat("\nRESULTS:\n")
if (selected_model==1)
{
    cat("\nThe best fitting model is y = [1/(a + x)] + c\n")

    A <- coef(non_linear.model_1)[1]
    B <- coef(non_linear.model_1)[2]
    C <- coef(non_linear.model_1)[3]

    y.model <- 1/(A + B*input.data$NaI) + C


    data.res <- data.frame(
    Measured_Iodine_Concetration=input.data$NaI,
    Measured_Fluorescence=input.data$Fluorescence,
    Model_Fluorescence=y.model)

    writeWorksheetToFile("Model.xlsx",
    data =data.res,
    sheet = "data_and_model",
    header = TRUE,
    clearSheets = TRUE)

    writeWorksheetToFile("Model.xlsx",
    data =tidy_model_1 ,
    sheet = "summary",
    header = TRUE,
    clearSheets = TRUE)
```

```
png("Fit.png")
plot(input.data$NaI, input.data$Fluorescence,
    xlab="Iodine concentration", ylab="Fluorescence",
    main="Hyperbolic fit", cex.main=3, cex.lab=2.5, cex.axis=2.5)

lines(input.data$NaI, y.model, col="red", lwd=2)
dev.off()

cat("\n\nDo you want to estimate the iodine
    concentration from florescence values? (Y/N) ")
continue <- readLines(file("stdin"),1)

if (continue == "Y")
{
    cat("\n\nType the name of the file
        with the new fluoescence's values: ")

    new.data.file <- readLines(file("stdin"),1)

    # new fluorecence data
    new.data <- (read.xlsx(new.data.file, 1))

    # iodine concentration
    x.model <- 1/(new.data[,1] * B) + A/B

    res <- data.frame(Fluorecence = new.data[,1],
    Iodine_concentration = x.model)

    writeWorksheetToFile("Model.xlsx",
    data = res,
    sheet = "Iodine_Prediction",
    header = TRUE,
    clearSheets = TRUE)

    cat("\n************************************************** \n")
    cat( " ***        The program terminates correctly.      *** \n")
    cat( " *** Results have been saved in Model.xlsx file. *** ")
    cat("\n**************************************************")
    cat("\n\n")

}
else {
    cat("\n************************************************** \n")
    cat( " ***        The program terminates correctly.      *** \n")
    cat( " *** Results have been saved in Model.xlsx file. *** \n")
    cat("\n**************************************************")
    cat("\n\n")
}

}

if (selected_model==2)
{
    cat("\nThe best fitting model is y = exp(a*x) + b\n")
    cat("Coefficients")
    #print(tidy(non_linear.model_2))
    #cat("\n")
```

```r
A <- coef(non_linear.model_2)[1]
B <- coef(non_linear.model_2)[2]

y.model <- exp(A*input.data$NaI) + B

data.res <- data.frame(
Measured_Iodine_Concetration=input.data$NaI,
Measured_Fluorescence=input.data$Fluorescence,
Model_Fluorescence=y.model)


writeWorksheetToFile("Model.xlsx",
data =data.res,
sheet = "data_and_model",
header = TRUE,
clearSheets = TRUE)

writeWorksheetToFile("Model.xlsx",
data =tidy_model_2,
sheet = "summary",
header = TRUE,
clearSheets = TRUE)


png("Fit.png")
plot(input.data$NaI, input.data$Fluorescence,
    xlab="Iodine concentration", ylab="Fluorescence",
    main="Exponential fit", cex.main=3, cex.lab=2.5, cex.axis=2.5)

lines(input.data$NaI, y.model, col="red", lwd=2)
dev.off()

cat("\n\nDo you want to estimate the iodine
    concentration from florescence values? (Y/N) ")
continue <- readLines(file("stdin"),1)

if (continue == "Y")
{
    cat("\n\nType the name of the file
        with the new fluoescence's values: ")

    new.data.file <- readLines(file("stdin"),1)

    # new fluorecence data
    new.data <- as.data.frame
        (read.xlsx(new.data.file, header=T, sheet=1))

    # iodine concentration
    x.model <- (1/A)*log(input.data$NaI - B)

    res <- data.frame(Fluorecence = new.data[,1],
    Iodine_concentration = x.model)

    writeWorksheetToFile("Model.xlsx",
    data = res,
    sheet = "Iodine_Prediction",
    header = TRUE,
    clearSheets = TRUE)
```

```
        cat("\n*************************************************** \n")
        cat(" ***        The program terminates correctly.    *** \n")
        cat(" *** Results have been saved in Model.xlsx file. *** \n")
        cat("\n***************************************************")
        cat("\n\n")
    }

    else {
        cat("\n*************************************************** \n")
        cat(" ***        The program terminates correctly.    *** \n")
        cat(" *** Results have been saved in Model.xlsx file. *** \n")
        cat("\n***************************************************")
        cat("\n\n")
    }
}

if (selected_model==3)
{
    cat("\nThe best fitting model is a polynomial of degree:")
    cat(paste(" ", degree, "\n", sep=""))

    A <- tidy(non_linear.model_3)[,2]

    term <- array(0, length(A))
    term[1] <- A[1]
    for (i in 2:length(A))
    {
        term[i] <- A[1]*input.data$NaI^i
    }
    y.model <- sum(term)

    data.res <- data.frame(
    Measured_Iodine_Concetration=input.data$NaI,
    Measured_Fluorescence=input.data$Fluorescence,
    Model_Fluorescence=y.model)

    writeWorksheetToFile("Model.xlsx",
    data =data.res,
    sheet = "data_and_model",
    header = TRUE,
    clearSheets = TRUE)

    writeWorksheetToFile("Model.xlsx",
    data =tidy_model_3,
    sheet = "summary",
    header = TRUE,
    clearSheets = TRUE)

    png("Fit.png")
    plot(input.data$NaI, input.data$Fluorescence,
         xlab="Iodine concentration", ylab="Fluorescence",
         main="Polynial fit", cex.main=3, cex.lab=2.5, cex.axis=2.5)

    lines(input.data$NaI, y.model, col="red", lwd=2)
    dev.off()

    cat("\n\nDo you want to estimate
         the iodine concentration from florescence values? (Y/N) ")
```

```
    continue <- readLines(file("stdin"),1)

    if (continue == "Y")
    {
        cat("\n\nType the name of the file
            with the new fluoescence's values: ")

        new.data.file <- readLines(file("stdin"),1)

        # new fluorecence data
        new.data <- as.data.frame
                    (read.xlsx(new.data.file, header=T, sheet=1))

        # iodine concentration

        x.model <- list()
        length(x.model) <- dim(new.data)[1]

        for (i in 1:dim(new.data)[1])
        {
            A0 <- A[1] - new.data[i,1]
            x.model[[i]] <- polyroot(c(A0,A[2:length(A)]))
            cat(paste("Solutions (", i, "): "),
                x.model[[i]],"\n",sep="\t")
        }
        cat("\n*************************************************** \n")
        cat( " ***       The program terminates correctly.       *** \n")
        cat( " *** Results have been saved in Model.xlsx file. *** \n")
        cat("\n***************************************************")
        cat("\n\n")
    }
    else{
        cat("\n*************************************************** \n")
        cat( " ***       The program terminates correctly.       *** \n")
        cat( " *** Results have been saved in Model.xlsx file. *** \n")
        cat("\n***************************************************")
        cat("\n\n")
    }
    }

}

process_fluorecence(opt$file)
```

5.3   CODE 3: Bi-exponential Regression

The following code implements the method proposed by Jacquelin [1] to estimate the parameters of a bi-exponential form.

Jacquelin [1] showed that a suitable integral equation can turn a difficult non-linear regression problem into a simple linear regression. The Jacquelin's method is based on the following principles of the linearization of the differential and/or integral equations. Given n experimental time points (t_i, y_i) located in proximity of a representative function $y(\theta; t)$, where $\theta = (a, b, c, \ldots)$ is the vector of parameters

$a, b, c, \ldots$, and t, in our study, is the time variable, and another function $f(t)$, we can approximate the integral and the double integral of the product $y(u)f(u)$ in the following way:

$$\int_{t_1}^{t} f(u)g(u)du \approx S_i, \tag{5.1}$$

where

$$S_i = \begin{cases} 0 & \text{if } i = 1 \\ S_{i-1} + \frac{1}{2}(f_i y_i + y_{i-1} f_{i-1})(t_i - t_{i-1}) & i = 2, \ldots, n \end{cases}, \tag{5.2}$$

where $S_i = S(t_i)$, $f_i = f(t_i)$ and similarly $y_i = y(t_i)$. with $i = 1, 2, \ldots, n$.
The double integral of $f(u)y(u)$ is thus

$$\int_{t_1}^{t} \left[\int_{t_1}^{v} f(u)y(u)du \right] dv \approx SS_i, \tag{5.3}$$

where

$$SS_i = \begin{cases} 0 & \text{if } i = 1 \\ SS_{i-1} + \frac{1}{2}(S_i + S_{i-1})(t_i - t_{i-1}) & i = 2, \ldots, n \end{cases}, \tag{5.4}$$

where $SS_i \equiv SS(t_i)$. Consider now a bi-exponential model for $y(\theta; t)$, with $\theta = (a, b, c, p, q,)$, i.e.

$$y(\theta; t) = a + be^{pt} + ce^{qt}. \tag{5.5}$$

Applying two successive integrations to $y(\theta; u)$ (for the sake of brevity denoted as $y(u)$), the function y can be expressed as follows [1]:

$$y(\theta; t) = -pq \int_{t_1}^{t} \left(\int_{t_1}^{v} y(u)du \right) dv + (p+q) \int_{t_1}^{t} y(u)du + Ct + Dx, \tag{5.6}$$

where C and D are constants depending on the lower bound of the integral. Equation (5.6) can be re-written as

$$y(\theta; t) = C_1 \cdot SS(t) + C_2 \cdot S(t) + C_3 t + C_4, \tag{5.7}$$

that is, a linear equation in the parameters A, B, C, D, and

$$C_1 = pq; \quad C_2 = -(p+q) \tag{5.8}$$

$$S(t) \int_{t_1}^{t} y(u)du \tag{5.9}$$

$$SS(t) = \int_{t_1}^{t} \left(\int_{t_1}^{v} f(u)y(u)du \right) dv, \tag{5.10}$$

where $S(t)$ and $SS(t)$ are estimated via numerical integration as formulae (5.2) and (5.4). Using Eq. (5.7), a non-linear regression is turned into a linear regression. Denoting with $\hat{C}_1$ and $\hat{C}_2$ the estimates of C_1 and C_2 obtained from the regression of Eq. (5.7), the estimates of p and q, denoted, respectively, with $\hat{p}$ and $\hat{q}$ are

$$\hat{p} = \frac{1}{2}\left(\hat{C}_2 + \sqrt{\hat{C}_2^2 + 4\hat{C}_1}\right) \tag{5.11}$$

$$\hat{q} = \frac{1}{2}\left(\hat{C}_2 - \sqrt{\hat{C}_2^2 + 4\hat{C}_1}\right) \tag{5.12}$$

```
###########################################################
#   Jacqueline method for fitting bi-exponential models: #
#   y(t) = a + b * exp(p*t) + c * exp(q*t)               #
###########################################################

biexponential.fit <- function(x, y)
{
    print("Bi-exponential model: y(x) = a + b*exp(p*x) + c*exp(q*x)")
    n <- length(y)
    S <- array(0, n)
    SS <- array(0, n)

    S[1] <-   0
    SS[1] <- 0

    for (k in 2:n)
    {
        S[k] <- S[k-1] + 0.5 * (y[k] + y[k-1]) *(x[k] - x[k-1])
        SS[k]<- SS[k-1] + 0.5 * (S[k] + S[k-1]) *(x[k] - x[k-1])
    }

    A.row.1 <- c(sum((SS)^2), sum(SS*S), sum(SS*x^2), sum(SS*x), sum(SS))
    A.row.2 <- c(sum(SS*S), sum((S)^2), sum(S*x^2),  sum(S*x), sum(S))
    A.row.3 <- c(sum(SS*x^2), sum(S*x^2), sum(x^4), sum(x^3), sum(x^2))
    A.row.4 <- c(sum(SS*x), sum(S*x), sum(x^3), sum(x^2), sum(x))
    A.row.5 <- c(sum(SS), sum(S), sum(x^2), sum(x), n)

    A <- rbind(A.row.1, A.row.2, A.row.3, A.row.4, A.row.5)

    B <- (c(sum(SS*y), sum(S*y), sum((y*x^2)), sum(x*y), sum(y)))

    C <- solve(A) %*% B

    p <- 0.5*(C[2] + sqrt(C[2]^2 + 4*C[1]))
    q <- 0.5*(C[2] - sqrt(C[2]^2 + 4*C[1]))

    beta <- exp(p*x)
    eta <- exp(q*x)

    A1 <- rbind(c(n, sum(beta), sum(eta)),
    c(sum(beta), sum(beta^2), sum(beta*eta)),
    c(sum(eta), sum(beta*eta), sum(eta^2)))

    B1 <- c(sum(y), sum(beta*y), sum(eta*y))
```

```
      abc <- solve(A1) %*% B1

      a <- abc[1]
      b <- abc[2]
      c <- abc[3]

      res <- c(p, q, a, b, c)
      names(res) <- c("p", "q", "a", "b", "c")

      y.model <- res["a"] + res["b"]*exp(res["p"]*x)
                 + res["c"]*exp(res["q"]*x)
      deviance <- sum((y - y.model)^2)

      res.final <- c(p, q, a, b, c, deviance)
      names(res.final) <- c("p", "q", "a", "b", "c", "deviance")
      res.final
}
```

Reference

1. Régressions et équations intégrales. https://it.scribd.com/doc/14674814/Regressions-et-equations-integrales. Accessed 02 Nov 2019.

Index

© The Author(s), under exclusive license to Springer Nature Switzerland AG 2020
P. Lecca, *Identifiability and Regression Analysis of Biological Systems Models*,
SpringerBriefs in Statistics, https://doi.org/10.1007/978-3-030-41255-5

Printed in the United States
By Bookmasters